Back in the Wild

by the same author

The Spider's Tales

Too Short a Day

Life with Daktari

Listen to the Wild

In the Wild

BACK IN THE WILD

SUE HART
Illustrated by LEIGH VOIGT

WITH A FOREWORD BY

GEORGE ADAMSON

COLLINS and HARVILL PRESS
London and Johannesburg, 1977

First published in book form in 1977

ISBN 0 00 262049-9

Set in Imprint

Made and printed in Great Britain by
William Collins Sons & Co. Ltd, Glasgow
for Collins, St James's Place and
Harvill Press, 30a Pavilion Road
London SW1

We dedicate this book to all
who endeavour to keep the wilderness alive

Acknowledgements

The articles published in this book first appeared as weekly features in the *Rand Daily Mail* during 1974, 1975 and 1976. I wish to express my thanks to the Editor for his kind permission to reproduce them in book form.

I would also like to extend my grateful thanks to the following for their encouragement, advice and constant assistance: National and Natal Parks Board; The Division of Nature Conservation of the Transvaal and the Cape; the Transvaal Museum; the Departments of Zoology and Mammal Research of Pretoria University; and to those who gave time and energy in assisting me to gather and assemble a great variety of material, especially wildlife scientists, rangers and conservationists who have so generously allowed me to share their successes and some of their failures, too, and have permitted me to take a close look into their world.

SUE HART

Contents

Illustrations

Foreword

Many people can write appealingly and with knowledge about wild animals in an impersonal kind of way, but very few have experienced and can convey the intimate feeling and compassion which Sue Hart shows. To her any animal, irrespective of size or value in human terms, is a living being, deserving the same care and treatment as does the human species.

Always, I shall remember with gratitude the never-failing response of Sue Hart and Toni Harthoorn to my calls for help to treat my sick and injured lions, often at considerable inconvenience to themselves, requiring long hours of driving over rough roads and bush tracks to remote places. And at the end, as happened on three occasions, performing intricate operations on full-grown lions, lasting several hours, in the field and under the most primitive conditions.

Back in the Wild is most fortunate in having the co-operation of Leigh Voigt, whose beautiful drawings so perfectly support the text.

May Sue Hart long continue to fight for the most noble and worthy of causes, which is vital for the quality of life on our planet.

George Adamson
'Camp of the Lions'
Kora, May 1976

Introduction

However often you journey into the wild, each experience brings new impressions, new kinships and renewed astonishment at the marvellous beauty and intricate pattern of Nature's realm. As you listen to the night under a bushveld sky or lie hidden in the sun-scorched grass, the dynamic, vibrant world around you engulfs and enchants so that you long to return again and again, as I did, to search, to listen and to absorb.

Here, then, is a touch of the wilderness, the merest fragment of the diversity, the magnificence and the mystery that is yet to be discovered, surely an intricate part of our innermost soul.

Sue Hart

Cheetah in Jeopardy

Eons ago the cheetah, often called the greyhound of the cats, diverged from its feline relatives by acquiring adaptations solely for its coursing method of hunting. Much of the specific equipment of the stealth-hunting wild cat was lost in the process, such as retractile claws and powerful jaws.

By the time the modern spotted sphinx emerged, it had gained unique status as the fleetest of world predators, a rank which, to date, has never been challenged though it could yet be, should prevailing conditions call for an even faster, more light-footed hunter of the plains.

But is this likely to happen? Will there ever be sufficient space for such a creature? It is improbable because the vast open plains grow yearly less verdant and plentiful.

Already we are fighting for the survival of this superbly-structured, enigmatic cat that can run at over 100 km per hour, kill instantly, feed faster than any other hunter (to protect itself from predators) then merge back into its special world, the dull pale yellow body blending perfectly with the natural dry grass-cover it needs for its very existence.

Take a look at it again when you can, even in a book if you cannot go out and find one . . . and they are hard to find. Perhaps only 2,000 remain in the whole of Africa – mostly in the Kalahari National Park, the Kruger National Park, game farms, small sanctuaries, zoos.

Recently, alone in my car, I had the luck to see a wild cheetah in the early morning. I counted myself very fortunate as I watched the supple body, gentle eyes, the passionate tear-streak lines to the mouth, the powerful long tail with its spectacular banded end.

The cheetah gazed at me, her swelling teats evidence of her sex. We looked each other in the eye, yet even then her tail moved slowly, from side to side, as if my presence provoked uncertainty, as though she felt impatience at my intrusion.

I looked at the cheetah's coat of solid spots (someone once counted 2,000 on one adult) and wondered at her hollow flanks, her hungry look and her air of patient waiting. Were her cubs alive, and if so, had she hidden them in some thick scrub out of danger's reach? Was she planning a hunt, was she resting between sprints, was she mourning the loss of her young?

Sitting dog-fashion on the slope of an old ant-heap, the cheetah had become alert, her small ears attuned to receive the slightest vibration, her sensitive nostrils alive to every scent.

Just ahead a dung beetle was engaged in a single-handed house-moving operation, rolling its castle towards the cheetah look-out. A limpid-eyed bush duiker suddenly appeared out of nowhere, took in the situation at a glance and dived back into the thicket from whence it had come, obviously unnerved by the cheetah's fixed stare in its direction.

I turned my head away to gain a better view of the flashing cinnamon wings of a Burchell's coucal. When I looked back, the elegant cat had gone, without even a rustle, or the swaying of grass stems.

There was emptiness round the ant-heap now; just brown earth where, minutes before, the lithe body had lit up the clearing with a contained vitality and vibrant beauty that only the cheetah, of all the streamlined hunters of the wilderness, knows how to project.

Thank heaven for rescue ventures, such as The Endangered Wildlife Trust, newly set up within Pretoria University's Department of Wildlife Management, which is barely three years old. It promotes the greatly-needed study of the cheetah's status.

The spotted cat remains secretive, baffling and vulnerable to change. All over Africa its numbers are dwindling at a most alarming rate and still only a fraction of what needs to be done to ensure its future is being achieved. And while this state of emergency endures, poaching and hunting continue for no reason but the value of the skin, which is soft and ornate.

'Is it too late for the cheetah?' asks Clive Walker, President of The Endangered Wildlife Trust, as he appeals for funds, this time through artist Paul Bosman's contribution of a superb painting and prints of the Cape hunting dog on the move, the latter being another species of wild animal threatened with extinction.

Soul of the Elephant

It takes years, many long years of bush apprenticeship before even an experienced lover of the wilderness can claim to be able to 'read' the elusive but very significant signs of the bush.

Baron Werner von Alvensleben, until recently director of the largest wildlife concession in Africa – a remote, untamed section of Mozambique territory north and south of the Save River – believes that the greatest tracker of the African continent (and the greatest hunter) has always been and will always be the African himself.

In an interview for Radio South Africa's transcription service series, 'Custodians of the Wild', the tall, tough, dynamic bush enthusiast spoke of this land, its people, and of the wealth of game in an area which is so vast and varied as to take your breath away. From the 2,500-metres-high Chimanimani mountains on the eastern Rhodesian border, it ranges down to the shores of the Indian Ocean, thousands of square kilometres of hauntingly beautiful yet tragically abused wilderness.

Werner von Alvensleben, an almost legendary figure in wildlife today, cannot speak highly enough of the consummate skill of his Shangaan assistants, without whom the tracks of game wounded by unseasoned hunters could never have been discovered.

Once a professional ivory hunter in Mozambique, von Alvensleben spent years hot on the heels of the greatest terrestrial mammal we know – the elephant. These years endowed him with reverence, respect and love for nature which comes to those – so different from the undisciplined trigger-happy intruders – who hunt with compassion and not with greed.

Having lived and thrived off the bush, von Alvensleben became an ardent game conservationist, determined to implement and uphold the strictest jungle justice ever seen in Africa. Is it possible to imagine the hours, days and months of agony inflicted by the twisted strands of cable in a snare – 170,000 brought to light during his time in Mozambique? Truly the curse of Africa.

For sixteen years the battle against illicit poachers, Whites and Africans alike, raged without respite. Gradually the number of nooses diminished by one third. The word spread, and respect and fear for the new preservation policy grew as the years advanced. By the end of 1973 it was estimated that a million wild animals had been saved from death.

Everything from the diminutive, enchanting Suni antelope

to the magnificent elephant had been and still is, though to a slightly lesser extent, being victimised. When snares are not regularly patrolled by the poachers, the trapped creatures are subjected to the vilest kind of horrors, including the loss of their eyes to vultures, pecked out while they are still alive.

Perhaps von Alvensleben's greatest empathy lies with the tuskers, 'the only animals that have the ability to reason'. To follow them, and 'read' their spoor, requires the deepest kind of kinship with and understanding of the ways of the wild.

What kind of ivory is he carrying – curving outwards? thick? heavy? What kind of temperament does he possess – 'spooky', sensitive to the point of carrying out a double-bluff by back-tracking his own spoor? Is he wary, fearful or confident?

Here is the three-toed with a hint of the four-toed imprint. He moves along browsing, comes to a water-hole, rests in the shade of an old ant-heap (on which he leaves an excellent tusk replica), yet when he moves on, he passes his favourite food source, the marula tree, with no more than a hasty branch-tug, moving on and on, now at speed. Why?

Somewhere along the elephant track another bull has passed across the trail, leaving the scent of his glands as an alarm signal for the next passer-by. Our peaceful, ambling elephant got the message and ran, full speed ahead, his instinct and memory, conscious or inherited, filling him with the dread of predatory man, his only enemy.

'They move along like a convoy,' said von Alvensleben as he remembered the great herds of the Save River basin. 'Their speed is that of the slowest member of the group, a new-born infant, cared for by the fighting bulls, the sentinels, the old and young alike. Here is a social unit, a really close-knit family. What a tragedy that this noble animal can never forget its fear or lose its suspicion of man.

'I am still learning about them. Their habits vary from one place to another; we know that.'

If only time and circumstance were more favourable so that Werner von Alvensleben and others could continue what they have begun without interruption from the turbulent, uncivilised outside world.

Insect Antics

Among wildlife researchers in southern Africa the Prozesky husband and wife team – he ornithologist and wildlife education officer, she an entomologist – is as remarkable as it is productive.

'Just five minutes, please, before you begin,' pleaded Proz (as he is usually called), entering his wife's Transvaal Museum office just as I opened my notebook to record the story of her current research.

I was delighted to see him, especially as he proceeded to project the most astounding series of sounds, ranging from a dramatic drum-beat to high-pitched pulsing oscillations. I recognised the rhythmic tap of the belly-beating tok-tokkie beetle, though until that moment I had not realised that there are so many sound variations on the same theme, transmitted by the same group of creatures. Among the *Tenebrionidae*, a genus of non-vocal communicators, each kind of beetle uses a distinct and very discernible come-hither mating call, male to female, consistently using the lowest extremity of the abdomen to effect vibration.

Vibration, acoustic communication, these aspects of behaviour are now being studied in ever-increasing detail.

With the help of the acoustic department of the CSIR (Council for Scientific and Industrial Research) Dr Lili Prozesky is gleaning results in her definition of insect characteristics and behaviour traits through sound analysis.

On her desk lay recently developed photographs of a remarkable creature known as the nocturnal tree cricket. It belongs to the genus *Oecanthus*, species *Burmeisteri* and is endowed with a genius for acoustic engineering.

Only 15 mm long, this orange-tinged, transparent-winged cricket has evolved his communication system to the point of perfection. About the time he feels the urge to reproduce, he searches out a large, soft leaf and gnaws a pear-shaped orifice into the centre of the blade. Having completed the first step he moves into his cavity with forelegs and anterior body section. Once he has comfortably settled, his lustrous, beautifully designed wings come into play. The inner surface of the left wing scraper rubs against the right wing file, using the leaf-edge as a baffle-board for more effective, longer-ranging sound transmission.

Without the aid of this leaf amplifying system his sound waves would become dissipated before reaching his mate. One chirp a second at 2,000 cycles of pure-tone vibration, on and on all through the night until the moment of triumph.

How does his lady receive the message and respond instantly to the summons of her mate without ears in her head?

Just to be different, she carries her 'ears' on her legs, one on each front limb. When she reaches her destination she mounts upon the cricket's back, proceeds to suck his dorsal gland and remains there while his sperm is transferred to her oviducts. She then dismounts and seeks a tree-stem in which to deposit her eggs.

Now she must bring her elaborate, spiny ovipositor into play and literally saw a hole into the stem that will accommodate her eggs, and protect them from harm, until the new generation of remarkable creatures is ready to launch out into their world.

Timeless Wilderness

We had almost reached the end of the trail, just fifteen minutes' walk from base-camp, clambering up the last slope which separated us from the river's rising chill. Our footsteps were flagging a little as we scrambled amongst the drapery of tangle-thorn, our single-file procession engulfed in the atmosphere of eventide that filtered from the bushveld wilderness: the scent of potato-plant mingling with that of wild gardenia, released in the cool dampness that follows the sun-sting of a sweltering African day.

There was the aroma of buffalo where scores of hooves had passed by a few hours before. Crested guinea-fowl were rasping themselves to sleep as the night-jar's lament and the chorus of nocturnal insects heralded the coming of darkness.

At the very last ascent our leader, John Varty, had hesitated

for a moment, then halted. As we topped the rise, the full red orb of the newly risen moon hovered in our path, seemingly within reach. Quite suddenly the glowing circle grew paler and rose up from the horizon of leaf-crowns like a balloon.

I looked back at the young faces behind me and saw awe and amazement playing across their features. There was nothing one could say, except to wonder and to give thanks. Ten of us, together in the wilderness, discovering or rediscovering its meaning and magnificence.

Of our party on the Wilderness Leadership Trail, the first ever to be conducted within the Transvaal (initiated by Ian Player in Natal), six were teenage scholars chosen on the strength of projects prepared for the symposium on Conservation and

Environmental Awareness. Sponsored by the Rotary Club of Durban, it had been set in motion by the Wilderness Education Trust of Johannesburg, on this occasion led by artist-conservationist Clive Walker at Londolozi Reserve, west of the Kruger National Park.

What a wonderful way to spend ten days, days which would probably leave an indelible impression on them for the rest of their lives.

A herd of buffalo, 37 with calves, were feeding noisily along the swamp reed bed of the Sand River, trampling everything about them as if they had no care in the world. Then suddenly, very near them, there was a flash of silken sheen and a leopard slid momentarily into sight, just below the rock clump look-out which commanded a view of the whole terrain.

In an open expanse the buffalo cow would have turned and repelled any aggressor but here, amongst dense riverine vegetation, it was much easier for a calf to be cut off from its dam and snatched out of reach.

We heard an adult bellow, then the lowing bleat of a calf which was lost as a bubbling coucal reached its crescendo.

Warthogs were resting under the fig-crowns, while three kudu bulls with immense horns moved through the bush with unhampered, easy grace. Nearby a pair of steenbok kept close company, the buck sleek and muscled, the doe hornless, her face possessed of an ultra-feminine sweetness.

There was much to take in, to absorb, and to discuss at the end of each enthralling day as we looked into the glowing embers of the camp-fire that seemed to epitomise the freedom of the wild. During the last hour, before we cocooned ourselves into our sleeping-bags set under the open sky above the river, I asked the Wilderness Trail gang to think aloud into my microphone.

'A remodelling of my values, a place that provides peace and solitude and is, oh, so utterly beautiful,' said Diane du Toit, aged 17. She could not wait to pass on what she had just discovered.

'For me,' murmured Margaret Parker, 'the wilderness means not being held down by time or confined by boundaries. It means great expanses, open spaces . . .'

Eyes of an Owl

A wise old owl sat in an oak,
the more he saw the less he spoke,
the less he spoke the more he heard,
why can't we all be like that wise old bird?

Why not indeed, I thought, waiting for the silent, down-clad hunters to return, my eyes stinging with weariness as the moments ticked by.

Two hours before midnight I must have dropped off, for the

sound high in the trees jolted me out of my sleepy state.

'Whooo – hooo – whoo – hooo.' How can one attempt to transcribe that mystical incantation into the written word? The call was low and throaty, echoing across the glade. The male parent owl was somewhere in the shadows, almost certainly with prey to satisfy his ravenous, month-old brood of two.

The female responded with a short burst of approval, her hoot a few tones higher in pitch. To my left, well concealed in his hessian hide on stilts, naturalist Howard Kirk switched on powerful arc-lights, their beams illuminating the branch-cavity nest in the old jacaranda tree which contained the enchanting owlet fledgelings. I had seen them several times earlier that evening: enormous eyes set in spectacular facial discs, perhaps acting as parabolic reflectors to catch light waves?

The owl has been endowed with an efficient sound receptor system of enormous survival value, since this twilight to darkness predator must live on rodents such as mice and rats whose high-pitched voices give little clue of their whereabouts.

The camera was whirring at the exact moment when the owl flew in, flitting with completely inaudible wing-beats, the velvet fringes set on the surface of the flight feathers acting as silencers. Attacks on insects, frogs, rodents must be swift and efficient. Sixty million years of owl survival (the first fossil record is in Wyoming, USA) bear witness to the success of their hunting method.

Prey in her beak, the russet-chested, seemingly enormous bird alighted on the nest stump, her vice-like claws releasing the rodent to the young, a fraction of what each owlet requires during one night. Her duty completed, she hesitated for an instant, turning her head towards the sound of the hidden camera. Through my binoculars I could see her with startling clarity, eyes like two shining orbs reflecting the bright arc-light. Most animals' eyes seem to be much less ostentatious, less visible within the camouflage pattern of the whole shape. At times they 'disappear' almost totally to make for easier protection or aggression. Were the owls' eyes created bright and large to evoke fear, even terror in their victims? In some insects, such as the

Cligula moth, the 'wings' warning' pattern most successfully resembles the eyes of an owl.

The lights were switched off. The parent bird flitted into the darkness to continue the hunt for food, or perhaps she was waiting for the male to bring his offerings to her, which in accordance with owl ritual she would give to the young.

Now there might be minutes or hours of waiting for the wood owls' return, but there was no dullness or boredom. The fruit bat's sonar location 'ping' signal filled the night. Frogs and nocturnal insects scraped and chirped their eternal melody.

Wood owls had first been sighted on this lowveld farm fifteen years back. Now, for five consecutive years, the same nest had been their base-camp at breeding time, though the tree stands but a few yards from the homestead. The owls remained shy, aggressive even, to anyone approaching the nest. That very night I had seen Howard's bush hat ripped off when he passed the tree. Even so, confidence that here they would not be harmed must have become deeply ingrained in their consciousness for on the Kirks' thriving farm, near White River in the Eastern Transvaal, agricultural activities and needs do not conflict with the well-being of its wild inhabitants, such as snakes, buck, insects and birds. Indeed the farm has become a recognised haven for the wild, lesser creatures, some of which, like the wood owl pair, have grown to depend on the farmer's kindly protection. In this way Howard not only maintains pockets of wilderness on his domain but constantly studies and learns new facts concerning his protégés, their needs, their communication methods, their territorial behaviour.

The 133 species of owls and their ways have always held a particular fascination for man. For centuries a symbol of life, as well as of death, a feature of medicinal and mystical belief, a forecaster of weather, mascot of the hunter, totem of wisdom throughout Greek and Roman civilisations, the owl is a segment of Nature's jigsaw, closely interwoven with man's own well-being in curtailing rodent populations.

It is the kind of bird one could easily spend a whole life observing.

Patterns of the Past

Dr Anton Rupert's magnificent plan to restore the Karoo's ecosystem, one of the oldest in the world – to turn back the clock so that, once again, we may witness herds of antelope on the vast plains – calls for a celebration.

I celebrated by turning back the clock a few million years in the company of that dynamic zoologist, the head of the Transvaal Museum's department of palaeontology, Mrs Elizabeth Vrba.

For some years now she has devoted a great deal of time and energy to the study of fossil antelope which once ranged over this land – antelope which were contemporaries of ape-man and early man.

The quality of early hominid environments has become a question of utmost importance to palaeontologists who seek clues among antelope and other fossils which provide some fascinating, though at present only tentative answers. These vegetarian animal fossils provide information about their role as prey animals of that time, for they formed part of early man's diet.

The early and mid-Pleistocene period must have seen a great variety of species, several probably bearing close resemblance to modern South African forms such as springbok, roan, reedbuck and kudu. Some, like the gazelle, have not been known to occur in southern Africa for hundreds of thousands of years. Others, such as one large fossil antelope whose nearest modern relative is probably the musk-ox of Greenland, would appear exotic and strange today.

Years of analysis of fossil remains have given Elizabeth Vrba a deep and new insight into possible ecological patterns surrounding ape-man and early man. Even where antelope fossils cannot be definitely identified as belonging to a particular species (thus giving a clue as to whether the environment was bush-covered, open, dry or wet) the physical features of the remains often reveal interesting facts: high-crowned teeth with large grinding areas tell of grazing rather than of the mastication of succulent leaves; resonating chambers of the ear (auditory bullae) suggest by their size whether or not the animal received sound waves best in thick or in open country; elongated limb bones tell of the speed essential to the inhabitant of the open savannah.

Professor John T. Robinson, formerly of the Transvaal Museum, went so far as to interpret the dentition and related features of the larger, robust ape-man as being indicative of a vegetarian diet. Another interesting idea which Mrs Vrba mentioned concerns the structure of the so-called pre-orbital fossa, a depression set anterior to the inner angle of the eye. In our modern antelope this contains a scent gland opening onto the facial surface which, at first glance, appears to be nothing more than yet another marking characteristic of the particular species.

But take a closer look, and watch an antelope – perhaps a blesbok – rubbing its face against the vegetation with a more than random movement, at times allowing a blade of grass to contact the external orifice. This ritual is now interpreted as territorial marking, a process undertaken with a patient care and gentleness which belie the fierce spirit immediately aroused by any unwelcome intruder who challenges the 'fence-line'.

Elizabeth Vrba's observations show that many early antelope forms possessed very much larger pre-orbital fossae than their descendants in whom the gland is reduced or absent. Why, one may well ask, should this pre-orbital gland be losing its importance? Is it because antelope, in general, have become less strictly territorial over the last few million years? Could it be that increased mobility offers selective advantage to antelope? And if so, how will this altered life-pattern affect breeding cycles, herd-harem protection, communal security and collective defence against predators?

The Incredible Honey-Bee

I am ashamed to confess that it isn't very often that, even when I am enjoying a spoonful of fragrant honey, I give a brief thought to the astonishing honey-bee, first described in 15,000 BC.

My son, Guy, who has spent weeks among those who work to bring in our supplies of honey, has often reminded me that bees are at least as fascinating and important as the more spectacular creatures I go so far to find. How right he is!

Discovering the world of bees is a never-ending adventure into a strange, near-magical dimension; here perfection of achievement and the finest division of disciplined labour are everyday events in the (queen-orientated) matriarchy of *Apis mellifica*.

Dave Falconer, a lowveld (White River) bee farmer of high repute, recently returned from an apiarist fact-finding journey to America where the honey-bee's role as pollinator and hence its top-ranking focus in agriculture is understood, recognised and acted upon.

Millions of dollars are spent on learning more about the bee from every angle: behaviour patterns, communications and adversities, especially the hazard of excess or ill-advised use of insecticides on nectar-carrying crops. These and many other avenues of research have become the concern of entire departments and university sections.

I was particularly entranced by a soundtrack of American bees (far less persecuted and hence less aggressive than the African species) which gave me an insight into their complex life-style, and highlighted their evolved communication system.

First, the sound of a contented swarm going about its business – a steady murmur of all's well in the world. In contrast, the higher-pitched, distressed buzzing of a swarm which has lost its queen, unquestionably recognisable as a panic signal. Next, the swarm in a state of anger, a very different tone which spells aggression. In this case a drop of chemical, its smell akin to that of an interloper queen bee, had been placed into the hive to evoke the reaction.

Two as yet unexplained sounds came next, the 'motor boat' call, very like a feline purr, then the 'piping' sound of the queen bee, produced by the rubbing of wings. Last of all, the amazing call (uncannily like the honking of Egyptian geese) of two queens in conflict. This must be heard to be believed.

I wonder if man is aware that these astonishing creatures which produce the only substance on earth which never deteriorates, build the strongest (for their size) structural unit known, the hexagon wax cell, and have evolved a highly-specialised nectar location ritual known as the bee-dance which may vary slightly in style and movement but is nevertheless always reliable.

In *The Insects* Peter Farb explains that when this 'tail-wagging' dance tells of nectar sources beyond the range of 100 metres, for instance, the worker bee makes a flat figure of eight with a straight line between the loops. This line acts as pointer, not at the flower nectar source but tells bees how far to the left or right of the sun to fly when they depart.

The 'language' of bees may seem incredible, yet it has been scientifically confirmed, as has the complex, short life-cycle (in busy times the honey-bee may literally wear itself out within five weeks or less) of the worker bee which begins life as nurse to its younger brothers and sisters but is soon promoted to hive cleaning and comb building.

Well equipped with eyes, antennae and dense body hairs to catch pollen, it finally leaves the home-hive for the arduous and dangerous task of pollen and nectar collection.

Dave Falconer, who takes good care that his honey-robbed

hives contain sufficient food for their occupants, produces 150 tons of honey each year, representing the journeys of . . . well, who can calculate the incredible, endless journeys that go to make up the honey-pot on your breakfast table?

The story of man's relationship with bees is vividly depicted in a rock-painting, dating back to the Palaeolithic era, discovered in Valencia, Spain, portraying two men climbing up long ropes, thought to have been woven out of sedge grass, to a small hole in a cliff representing the dwelling of a swarm of bees. One of the men is shown taking the honeycomb out of the cavity and placing it in a basket. Bees are seen flying around.

Thousands of years later the ancient Egyptians went in for migratory beekeeping up and down the Nile River. Hives were placed on rafts from which the bees flew to gather honey. Gradually the rafts were moved farther down the Nile to a point where there were more flowers.

From the first Dynasty of the Pharaohs to the Roman period the titles of kings were always associated with the sign of the bee, the cartouche containing the king's name being preceded by one. In Assyria (2050–1950 BC) the dead were coated with wax and buried in honey.

And so, from the time of King Solomon to that of Aristotle – the latter being the first man to deal with bees in a scientific manner – bees have been closely interwoven with civilisation. In Roman Law (100 BC) bees not enclosed in a hive were legally considered masterless.

'Bees are wild by nature,' states the law. 'Therefore bees that swarm upon your tree, until you have hived them, are no more considered to be your property than the birds which build their nests on your tree.'

It was much later, however, that the mystery of the queen bee and her intricate breeding cycle was solved by Janscha, the royal beekeeper to the Empress Maria Theresa of Austria. It was he who in 1771 discovered that mating occurs away from the hives.

We know that the queen is the most important personage in the hive and that she directs the course and success of her swarm, which is often composed of 50,000 individuals, each one occupied with a vital task.

Structurally, the queen is much like the worker, except that she is longer, and her reproductive organs are fully developed. The same egg which will produce a worker will also produce a queen. At times two queen bees develop simultaneously, causing acute conflict, since usually only one can hold sway over her colony. If one queen emerges from her cell before the other, she may gnaw a hole in the cell and sting her rival to death, or they may fight to the death, while the inhabitants of the hive look on. She does not lose her sting in the act of killing another queen.

The whole fascinating story of this subtle intelligence within the hive has yet to be unravelled. How does an emergent queen know that the embryo of another cell is a potential queen and not a worker?

There are times, however, when the queen loses favour and is mercilessly 'balled' out of existence by her own kind, that is stung to death by a dense 'ball' of bees, all struggling to get at her.

According to Dave Falconer even queens have their failings; if not wholly productive, as happens when old age or sickness overtakes them, they are no longer tolerated by the swarm, fated to meet a violent end in the midst of their own kind where, perhaps, a new queen is already taking over the marathon task of reproduction.

Have you ever seen a swarm on the move, clustering together in their thousands, one upon the other, somehow surviving pressure and density without apparent discomfort? What is the secret of their aerobatic link-up? Who holds up whom in such perfect order, without panic or stress?

Dave Falconer has become so bee-orientated that he often stops in the midst of one of his own problems to consider what the bee would do, in such an event, to solve it. He insists that we have a great deal to learn from bees, which are never idle, are always purposeful, sociable, communicative, totally selfless, and possessed of tireless energy.

Each substance, from honey, royal jelly, wax, pollen to space-filling propolis, is perfectly utilised, as a result of a superbly streamlined team effort which surely is something much more significant than an inherited pattern of life.

Rhino in the Mist

Thirteen years ago the white rhino cow known as Umfolanzana was captured in her native Zululand and successfully re-located in the Kruger National Park, an area of two million hectares.

Recently I saw her and her mates with District Ranger Thys Mostert, safe within a fence two metres high and extremely strong to prevent intrusion by carnivores. This terrain is a completely natural portion of the wild world outside it, and accidents can and do still occur. For instance, one female rhino had, previous to my visit, died of a twisted intestine, the first casualty in thirteen years, its tragic corpse as yet untouched by

hovering vultures which, for some reason, resisted the temptation of meat in this specific area.

Thys told me the strange tale of those aerial scavengers, which no one has been able to understand. The vultures circle down, perch on the trees, yet never descend to feed from a carcass as is their normal habit, an enigma which may take years of observation and research to resolve.

Great skill and patience are involved in effecting safe capture of a rhino and initial taming to a point where the presence and interference of human beings no longer evoke frenzy or aggression. Apart from this, there is the detailed preparation for the long journey of the captives which terminates in their release into a totally strange environment. The newcomer must acclimatise to new herbage, unfamiliar scents, strange bird calls, and to the loss of well known paths and waterways, trees and dung heaps that used to spell comfort and security.

Unless the man responsible understands his charge, long range re-location can set in motion a chain of detrimental physical and psychological effects. Ranger Mostert has the kind of experience and empathy with the new arrivals that spares them as much grief as possible.

Right up to their release from the large enclosure everything is conducted with the utmost care and forethought for the creatures that are being launched on their final journey into the vast bush-country.

I watched Umfolanzana's 18-month-old bull calf moving about with a ponderous yet strangely agile gait. He preceded his mother (black rhino calves always follow behind) as she moved along her special trail, undisturbed by our presence. Today this herd of white rhino can walk in peace in their new home, the end result of a heroic operation that had arduous beginnings and suffered almost insurmountable adversities.

The rhino herd have their own personal attendant: game scout Tsulea Moiana suddenly appeared out of the bush and gave his report of the day. All was well. He and his co-scout had patrolled and observed each animal within the fence-line noting their health, mood, behaviour.

Tsulea had little to fear from those large, lumbering, square-lipped grazers, so very different from their browsing, aggressive, hook-lipped 'black' counterparts. But in years to come the little bull might show a less placid side to his nature, for when the female comes into heat (oestrus) all hell is let loose if the bull-in-possession is challenged. Those long horns make lethal weapons which strike deep and violently as combatants rush at each other to the accompaniment of screams and roars.

According to Ian Player, famous pioneer of the white rhino re-location, the rhino's role in nature becomes especially significant when viewed in the light of his digestive waste.

The dung beetle rolls balls of dung to some special pre-appointed place to lay its eggs. Harvester termites carry away undigested grass stems to their nest. Banded mongoose, guinea-fowl francolins, rollers and shrikes feed and dust-bathe in the heaps, and even lizards have been seen to dig in the mounds of discarded vegetation.

How prehistoric these rhino appear when compared to the neatly-shod zebra and sleek, alert eland. From cave paintings we know that palaeolithic man killed rhino 25,000 years ago.

The Kruger National Park now supports about 400 white rhino, all imported, and a number of the black variety as well. Yet not many years back the battle for their survival had reached crisis level.

In his book, *The White Rhino Saga* (Collins, 1973), Ian Player writes of his first confrontation with white rhino, moments which were to change the course of his life:

'Two bulls loomed out of the mist. I had a perfect view of their physical characteristics as they walked along a ridge. The mouth was square and the nuchal hump between the head and the withers bulged prominently. Flies clung to their flanks and steam rose from their backs. These were truly creatures from a bygone age. The two rhinos grazed as they moved, their heads swinging in a scythe-like motion as they fed on the grass. I watched them move through the grey nthombothi trees into a cluster of candelabra aloes, and disappear into the mist.'

For the Love of an Egret

The African spring was very much in the air on the day I visited a flourishing lowveld reed-sited colony of breeding egrets and purple herons. It was a misty late afternoon, the grey sky a striking contrast to the sparkling white flight formation winging its way back to its home-base. Even the crested guinea-fowl (hardly a romantic bird) seemed to be in an amorous mood,

back-crests raised with emotion as the mating-game proceeded.

Down among the dense water-vegetation into which we pried with our binoculars, cormorants stretched their ample wings near the nesting sites of their loudly vocalising colony mates. They were rasping, honking, and yapping in the well-protected backwater of their farm sanctuary without an apparent care in the world. Their close inter-species existence provided protection to the 200-strong unit, protection against predators such as the leguaans, which take their eggs, and predatory birds that prefer to attack single, and hence more vulnerable, victims.

The purple heron is secretive and shy, moving in the shadows of dusk and dawn to feed on fish and insects. The egret, on the other hand, is an extrovert, easily visible bird whose life-style of seasonal change in colour and manner makes one wonder at the pattern and purpose of living things.

Doubtless, said naturalist Howard Kirk, my guide that day, everything is motivated by the urge to attract attention, to nest, breed and raise a new generation of lusty infants successfully. Courtship behaviour, such as is ascribed to demonstrative birds and other animals, has, in the past, often only been described as automatic, expressed in simple, scientific terms by students devoted to the study of behaviour. The well known British naturalist Maurice Burton, however, suggests that emotional qualities, strictly comparable to those more normally associated with human beings, are very likely to be in evidence, especially as courtship behaviour may be seen at times other than the breeding season, woven 'into the fabric of everyday behaviour'.

With human beings, says Burton, engagements may be short or long, love may be at first sight or only after a long period of 'being together'. In long engagements there may be bickerings and tiffs before a couple settle down. A close observer may well record all these traits in birds.

Among cattle egrets (*Bubulcus ibis*), the September onset of the breeding cycle heralds a fascinating, prolonged demonstration of the step by step ante-nuptial display and final consummation of a bond. Male birds find fixed sites and begin to act the part of the aggressive, hard-to-get mate. Their attitude, calls and demeanour attract the attention of the preening, absorbed spectator hens, who are no doubt enthralled by the sight of raised body feathers, erected plumes, and spectacular wing movements. At this time the egret's voice is dull and nasal, his abuse and beak violence acting like a magnet to the hen-party of admiring, eligible birds.

After days of territorial display on the part of the male, the female's demure, tentative approaches change to more definite advances, formalised by a show of reciprocal courtship which eventually settles into a mood of conjugal delight. By this time the whole colony is actively engaged in affectionate, progressive love-play, which after actual mating has taken place gives way to the remarkable ritual of stick collection, when the alighting bird brings a twig, gives it to the waiting bird at the nest, then flies off for another stick. The bird on the nest arranges the new twig in the most useful manner and is not satisfied until it fills exactly the right spot in the growing structure. This has been organised with the utmost care and skill and seemingly with a conscious sense of enjoyment in the performance of joint activity.

Perhaps one of the most amazing phenomena of egret life is that of plumage change throughout the season, involving very definite colour phases. In the 'off' season of autumn and winter, beak, feet and iris are yellow, dark green and pale yellow respectively, but in spring and summer at the onset of the breeding season they take on a startling coral-red hue. Plumes turn brown-buff in readiness for the sex-display when the egret voice alters its tone and frequency from quiet, husky unmusical notes to rasping, snorting, high-pitched repetitive calls.

What a world lies within those reeds, fulfilling a new cycle of life, a world that few would suspect, seeing the buffalo or cattle camp-followers feeding on the insects stirred up by the restless hooves.

At last, all was silent among the reeds, all except for the gay-striped frogs' chant and the final roosting call of the guinea-fowl above our heads. We retreated back to the car to save ourselves from being drenched by the rain we had been too absorbed to notice, until all sight of the heronry had been swallowed up by the starless, mist-laden evening.

Wild Heritage

Many creatures, great and small, from elephant to guinea-pig, Samango monkey to warthog, kudu to lion, just to name a few, came into sharp focus at Pretoria University's wildlife centre of zoology, where the Institute of Mammal Research had organised their sixth annual meeting on 1 November 1974.

Scientists came from as far afield as America. Many aspects of field-study and wide-range research (such as Marion Island's seal investigation) were reviewed and discussed intensively for three days.

Over 35 scientific papers were presented. One of the most dynamic of these contained the well-known ecologist Ken Tinley's plea for greater care of the base structure of our environment which is governed by the presence and flow of water at every level. Calling for a complete revision of certain vital ecological principles, he highlighted the urgent need for southern Africa's reassessment of landscape planning. In the long view this must be determined by the salient factors governing the whole ecosystem (defined as a system interrelating all strata of natural creation) rather than single, isolated areas artificially enclosed by, say, a fence-line.

Most current management programmes set up to protect natural systems, and especially endangered and rare species, are dealing with the effects, not the causes, of the changes. What is being protected may, under normal processes, be squeezed out and replaced by animals better adapted to natural environmental change.

Disturbance to one part of such a system can set up a chain reaction, the results of which are hard to imagine, let alone predict. Tinley maintains that unrestrained, ill-informed planning continues unnecessarily to damage unique natural systems containing rare plant and animal species. A sobering thought which, hopefully, will be noted and acted upon while there is yet time.

In total contrast to Ken Tinley's far-reaching views on environment, biologist Sonya Viljoen told of the life, love and communication of that most enchanting of our wilderness inhabitants, *Paraxerus cepapi*, the bush squirrel.

Sonya and I share an office at the university and so I was hardly surprised when, some months back, I found a warning of 'squirrels at large' pinned to our door. As I entered, I disturbed a mother and her new-born hairless pup (carried by the leg) busily engaged in investigating Sonya's precious manuscript-filled briefcase on a shelf from which the squirrel had no intention of being enticed to some other, more suitable nest site, such as the desk drawer. The squirrel was alert and nervous, yet totally in command, her movements fleet, her mood reflected in the constant flicker of her flowing tail.

In a remote flood-plain area of the Limpopo, Sonya had netted, collared, dyed and toe-clip-marked her study squirrels. She had recorded their dietary preferences, such as aloes (the leaves), wild berries, termites. Groups of up to 12 kept under constant observation had given her the opportunity to record their vocal communication and correlate this with tail response.

According to her findings, vocalisation is closely linked with mating activity, which lasts for about four hours. Contrary to many other creatures, it is the female of the species which lures the mate to her side. Sending forth an emotional sequence of whistles and clicking calls, in quality and tone very like the alarm signal, the female arouses and attracts the male who expresses his reaction through a soft nasal murmur increasing in frequency as the breeding season progresses.

Sonya recorded one female as producing six litters in a year (presumably a year of plenty), the pregnancy period of 55 days being 11 days longer than in any other known squirrel-type.

The calls she played at the Pretoria University conference ranged from the voice of a one-day-old – a just audible click – to the abusive, alarm sequences which sound like bird calls such as those of the Angola kingfisher. Squirrels call as they propel their seemingly weightless tail-angled bodies through space and into the safety of their holes, or into some leaf-thicket where, but for the flicking of their long wisps, they might never be detected . . . that is, if they could ever restrain their effervescent personalities and remain silent.

Voice of the Hyena

The last time I tried to catch up with biologist Simon Bearder's research into the life of *Crocuta crocuta*, the spotted hyena, he was virtually nose to the ground in examining spoor-prints in the bushveld.

Using the right front paw as his clue, he learned to relate the shape, length and breadth of the pad to an individual's weight, size, age, and condition, a difficult task when you consider that the hyena's drama of existence and its most active hours span from dusk to dawn.

Like a good detective Simon left no rock unturned within his study terrain, eventually gaining vital knowledge of the hyena's separate clans, their den sites, latrine areas, traditional pathways and the burrows to which they repeatedly return. He learned of their social behaviour, such as rivalry between neighbouring clans leading to display, fighting, and temporary association between group members, care of offspring and food hunting. Eventually this data will become part of a file of observations which will be interpreted in the light of modern game management programmes and practices set up for the ultimate preservation of any species.

At the 1974 Pretoria University wildlife conference Simon presented the sound picture of the much maligned but enormously useful predator-scavenger community living on the western border of the Kruger National Park within the Timbavati Private Nature Reserve. For over two years Simon and his wife, Catherine, spent hours recording the voices of the hyena, correlating a known individual's vocal expression with its age, status, sex and activity.

In assessing the significance of the recorded material a stranger-than-fiction array of sounds was projected, always in perfect synchronisation with visual examples.

'These characteristic calls of the bushveld may lose their mystery when analysed, but not their charm,' wrote Simon Bearder in his presentation summary.

First came the long-range 'whoops' – perhaps carrying information for individual recognition? Then the juvenile or infant high-pitched whoop, followed by the truly child-like food-begging cries of the young, a sound which, strangely, can continue into adulthood when it indicates submission, pecking order and clan discipline.

Simon continued with variations of the greeting moan, part of the inevitable ceremony of respect. Then the 'lowing' call, given in moments of danger, such as the proximity of a lion, a call which communicates intention to remain or flee. Next there was the 'bark' sound heard when food is disputed; 'rumbles' following aggressive encounters at a kill. Then the famous 'giggle' and 'cackle' when squabbling over a carcass and the low-pitched 'gruff' call, perhaps a warning to the young.

That we now know some of the meanings of these incredibly varied calls does not lessen the impression of impending tragedy or conjure up, any the less vividly, dark shapes that mete out death in the shadows.

The Masai tribe of East Africa tell a folk tale about the hyena, explaining how it came to be possessed of its very strange gait.

Once upon a time a crow met a hyena and said to him: 'You hyenas are always hungry, but you never complain. Why don't you climb up into the trees where the Masai store their meat and eat that?'

The hyena answered: 'Have we wings that we can fly up into the trees?'

Then the bird said: 'Go through the country and collect all the hyenas and I will tell you how to get the meat.'

So the hyena went off and journeyed far and wide collecting all the hyenas and eventually they all arrived at the crow's home.

Then the crow told them: 'First one of you goes and sits under the tree where the meat is. Then another one climbs on that one's head and a third one on the second one's head and a fourth on the third and so on and then I with my wings will fly up and guide the topmost one to the meat.'

So the first hyena went and sat under the tree where the meat was and, one after the other, all the others piled themselves on each other's heads and then the crow flew up beside them, but none of them got to the meat because the bird was tricking them and guiding them to the moon.

At last the hyena on the ground found that the weight of all the others was more than he could bear, so he wriggled out from underneath them to see what had happened to the meat. Then all the other hyenas came crashing down and as they hit the ground they broke their back legs, and ever since that day all hyenas limp.

A Tree for all Seasons

'We're five years old today,' said Koos Nel, curator of the Lowveld Botanical Gardens, set just outside Nelspruit, in the Eastern Transvaal of South Africa.

He was looking at his diary in which he had made notes for me over the past week – notes in answer to my query about trees and, in particular, the kind of tree that acts as bountiful provider for some of the birds that grace our land. As Koos realised the significance of the day, he reflected over the years since he had come to the lowveld and together we recalled the rambling wilderness that had existed in exactly the spot where we now sat in his comfortable office at the edge of the Gardens.

An area of 154 hectares of sanctified, protected floral wealth, at 640 metres above sea-level, it supports the widest possible range of plants within the marvellously varied habitat: rock formations in granite, kloofs, vleis and river banks – the latter shaded by great old trees such as the camdeboo (white stinkwood) and Transvaal teak – contrasting with the ferns, orchids and mosses which grow in the shade of their bowers. Flame-coloured aloes, succulents and the pride of de Kaap mingle with almost 500 species that occur naturally, in addition to those that have been introduced to provide a truly marvellous and complete display of what can and does grow in the lush, fertile lowveld.

The Botanical Gardens do not only provide visual gladness: one of its main roles is to collect and make available information about the care, handling, selection and preservation of flora. A great deal of research goes into botanical science, especially into the detailed study of such factors as plant growth, plant relationships, plant distribution – facts that are necessary for classification. The Gardens serve as a source of information for scientist, student and layman. No one is ever turned away, however inept their question may seem.

'How about a tree for all seasons?' I had asked Koos Nel at our first meeting. 'A tree of life that supports all kinds of animals, but especially the winged ones. A tree that provides food every month of the year.'

'That's a tall order,' the curator replied with a quixotic smile. 'But it so happens that there is a species of tree that will provide something for birds all the year round. If you're thinking of planting shade trees that are beautiful, and useful, that grow fast and are hardy, then take a look at the different kinds of fig trees. If you plan with wisdom, fig types will overlap the year round in producing fruit. Even in the coldest, driest winter seasons, the fruit eaters and those that take the insects within the fruit will never go hungry.'

Take *Ficus petersii*, the strangler fig, which in the wild germinates in the fork of another tree. It will support bird life from June until December and even January. In 1972 Koos observed such a tree to shed leaves in December yet to come into fruit through February into March!

Ficus sonderi came next on his list, bearing fruit at the same period as *Ficus soldanella*, the rock fig, from September until November.

The great sycamore fig bears its hairy, delicious fruit in the axils of its leaves, or in dense clusters along the main branches from December into February.

Ficus capensis, which grows up to 25 metres tall, had recently produced three crops of fruit in one year on a nearby White River farm, from January until May as well as during the other fruiting seasons.

Who needs all that fruit, and who would die for lack of it?

The list is long and impressive and includes such favourite birds as the tinker barbet, starlings, orioles, louries, the nocturnal fruit-bat, the sunbirds, the mouse birds, bulbuls, and the green pigeons which live almost totally upon the fig fruit.

As always in nature there are many variations on a theme. You might well find your tree bears fruit for longer periods or at different times, depending on soil, water, space, and other factors. This is a subject all of its own, as I soon realised when Koos began to explain the actual technique of transplanting and nurturing, which includes site selection, fertilising and recognition of stresses, especially in the early months and years of growth.

'If your tree doesn't look good it doesn't necessarily mean it is sick. Plants prefer certain localities. Some may react to drought or other adversities. Just keep trying and don't ever get impatient with a tree!'

There is no doubt that loving care and patience – and, of course, a great deal of knowledge – provide exactly the right kind of environment for the growing plant.

As Koos Nel proudly showed me some of his 2,000 young shrubs and trees I began to understand a little of what goes into his day and why it seems as if his vitality were linked to that of those well-cared-for thriving plants – endless neat rows of them.

He would not commit himself on the subject of plant-vibe transmission, reception or sensitivity, a vast subject which is, at present, gaining a great deal of interest amongst plant-orientated people throughout the world.

Elephant Frolic

'See that V-shaped tree stump ahead . . . and there, to the left of it a dead branch pointing upwards, and beyond that, under the bush, a brown patch that could be one end of a lion?'

All eyes, ears and noses – have you ever caught the whiff of lion? – were trained in one direction, excitedly waiting for the moment when the fawn blur would explode into the mane, face and flashing incisors of the king of beasts.

The December afternoon heat-haze has a way of playing tricks on one's already overtaxed vision, which makes it all the harder to detect elusive cats. Was that an ear flick, or the breeze stirring the grass heads? Was that a tail-lash, or a ground-bird gathering seeds? Surely those were amber eyes momentarily glowing among the wild flowers and surely that was a streak of white chin fur between silver-tinted leaves?

Minutes ticked by. Only sixty to go before the camp-gate would swing across the path, before the Christmas Eve sunset and deepening shadows would shroud the lions' terrain. Six whole days in the Kruger National Park, a wonderful way to spend Christmas away from the madding crowd and the strain of celebrations. With time on our hands, we could feel ourselves slipping into the rhythm and peace and pattern of the bushveld, sitting in one single spot all day if we chose to do so.

Looking down at the Olifants River in full flow, we had searched for the great tuskers among the baboon-inhabited, gold-glinting ancient sycamores, and discovered our first elephant busily engaged at his dust ablutions. His body hue almost matched the red-rust edge of the weather-worn cliff-face.

Gradually our attention was drawn to two young, adult tuskers whose movements revealed something more than the eternal quest for food. Interacting in the subtle manner of shy, enchanted lovers, they approached the river edge and entered the water, the male luring his lady step by step into the cool stream, his back to the current, until she succumbed to his charms and followed him in.

Not one of us had ever witnessed the courtship water-rites of the elephant, nor were we prepared for the sensitive interplay which now took place between them. Visibly entwined, they played and romped above water level, half-submerged, and below, blowing, touching, rolling, revelling, often so deeply under that only the end of their periscopic trunks gave a clue to their whereabouts – that and the swirl of water.

For one whole hour we marvelled at their joyful togetherness, wondering if, perhaps, they had mated below water and whether this was a prescribed manner of procreation when water levels are high.

Still moving nose to tail, they eventually returned to land and we were amazed to see another (probably a male) take to the

water with the same female, though in a much more restrained manner.

Was there something in the air? Impala rams sparring, zebras pairing on the plains, herds of sleek wildebeest with nurseries of calves, impala lambs romping and leaping in widening circles, speckled columns of giraffe of every shade from aristocratic deep brown-black to the light gold of the female in full milk.

Suddenly, without warning, my day-dream was shattered by the sight of an enormous leonine head and bronze mane rising out of the low bush. He yawned with deliberate arrogance and for some moments stretched his taut, muscled body in one rippling movement before lowering himself back into his private lair, where our prying eyes could not follow him.

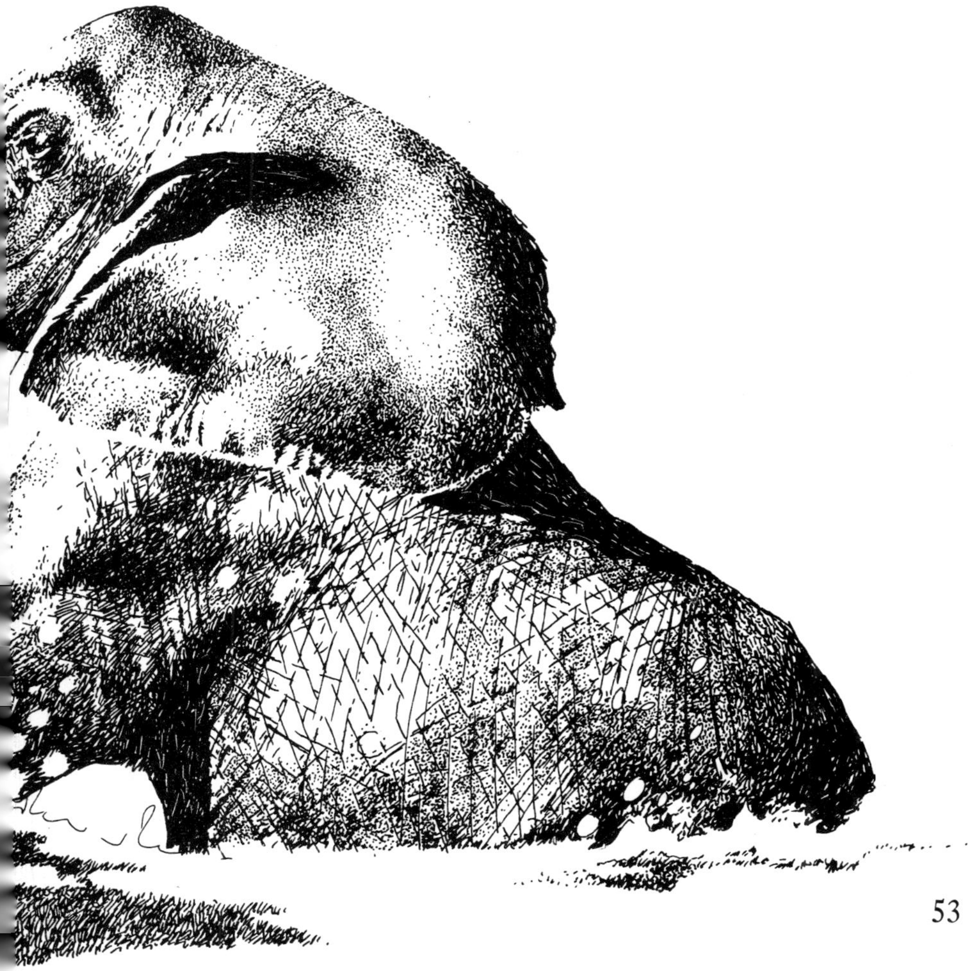

Kori Courtship

Exploring a wilderness area can be disappointing, however deeply you search for signs of life. You may sit at a waterhole through the hours of mid-day waiting for the bush inhabitants to come down, but there are times when the arena remains bleak and empty and your spirits drop as the minutes tick by.

Then a warthog appears, wallowing in the mud, and your morale lifts a little. But your companions have never been to a wildlife sanctuary before and are beginning to think that your tales of charging buffalo and mating elephants must be figments of your imagination. You try to draw their attention to passing birds but they show little interest.

On one well-remembered morning, just as I had run out of ideas and considered leaving for home, a station-wagon pulled up beside us, the occupants' binoculars riveted half-way up a marula tree which stood tall and slender in our line of vision. My co-travellers seemed suddenly to snap out of their discontent and take notice. Within moments they, too, were playing their binoculars on the marula, searching for the winged phenomenon which was causing such a stir.

'Look at this, isn't he glorious? Of everything we've seen today isn't that eagle the most beautiful?'

Bless the bird-watchers, I thought in relief; they had certainly injected new life into my crew.

Perched high up on a tree limb, the startling russet and black bird with bright red legs looked like the lord of winged creation. Perhaps alarmed by the stir he was causing the raptor suddenly took to the air, speeding on his way with a rocking, rolling motion, wings held like extended fingers, every inch the aerial acrobat from which he had long ago derived his name: bateleur.

My visitors scrambled for the bird books I always carry, and searched for the eagle; the bird-watch had begun in earnest!

'I had never thought', said one of my overseas trio, 'that looking at birds could be such fun.'

We began to look up birds we had seen that day which I had

fortunately noted along the way and I explained that interest in birds means there can never be a dull moment anywhere in wild country, at any time of the year. I told them of my last visit . . . when my own, limited knowledge had been augmented by a nature enthusiast with strong ornithological leanings who could not pass a single bird without stopping and trying to identify it. It had been late December when the prolific red-billed Quelea had been gathering in their thousands, weaving smoothly synchronised flight patterns in and out of the bush, in anticipation of the nesting season. We had identified tawny and martial eagles, secretary birds, lilac-breasted and European rollers in their flashing glory, the Marico sunbird, shrikes, kites, the hammerkop, the white stork on his migration, marabou, louries, herons, coucals undulating their rain-making call, hornbills, owls, barbets and there were others that we couldn't name.

It would be hard to say which of the birds afforded us the greatest visual pleasure, though perhaps I found the tall pair of Kori bustards mate-displaying in the long grass one of the most astounding sights of that week.

Strutting several metres in front of her male, the less ornate female appeared to give no sign of awareness, yet must surely have been affected by the Kori's courtship.

Well within our sight, but hardly within hers, the crested bird suddenly inflated his long neck with such fervour that the smooth neck-feathers momentarily became erected into a ruffle of grey-stippled plumes.

In a flash the demonstration was over, the neck deflated to normal size as the Kori once more pursued his mate at a respectful distance, leaving us to marvel at the courtier's ritual we had come upon so unexpectedly, a spectacle that few who come to Africa witness even once in a life-time.

Armada of Turtles

It was along a wild stretch of the Mozambique coast north of Lourenço Marques that I watched a marine turtle being massacred by holiday-makers for no reason but man's primitive urge to destroy. It was horrible. I remember feeling ashamed of being part of the human race and of having reacted too slowly to be able to halt the slaughter. For years afterwards that windswept, reef-horizoned coastline of turtle anguish would flash back at the oddest moments and I often wondered how many thousands of such innocents had met with a similar fate.

At last, in recent years, turtle conservation has taken on a new urgency and at least a portion of the public has begun to take an interest in the dramatic life-cycle of this amazing reptile, inhabitant of two worlds.

The largest of all marine turtles is known as the leatherback. Reaching up to 68 kg in weight, it is one of the sights of the seashore you are not likely to forget. If you are very fortunate, you may see the leatherback giant along the coastline of South Africa's Tongaland. The green turtle is the only reptile to migrate over long distances. It is known to have covered up to 2,240 km.

It was not until I was invited to see South African zoologist Dr David Hughes' remarkable film epic of the Ridley turtle that I began to understand something of the magic of its life. It is spent mostly in the open sea, that is until the female's urge to nest terrestrially drives her towards, and then on to, land.

While engaged in marine biology at the University of Miami, David and a colleague discovered two remote beaches in Costa Rica which the unique Ridley turtle has chosen as nest site. Unique because most of these turtles nest together on one single beach which stretches for 1,000 metres, the actual nesting site extending for less than half the distance. The only other five Ridley turtle beaches in the world, all in Mexico, had tragically lost their nesting nomads within a space of four years, victims of the Mexican taste for turtles.

David describes how on three or four nights each month vast

numbers of egg-laying turtles find their way to the Costa Rica observation beach, as many as 100,000 nesting together during the short laying span. When they return to the ocean as many as ten million eggs are left behind, all on that short stretch of 400 metres.

Watching this extraordinary phenomenon and making his first attempt to record it visually – rather than by means of the written word – drew David Hughes from his chosen scientific path to the competitive, less secure, and much more challenging field of camera presentation. Fortunate in his background of ecological training (he holds a doctorate) his filming gives a spectacular, in-depth view of the armada of turtles converging upon the shore, their clumsy progress on the sandy surface, their blind impulse towards a suitable nesting site, and their eventual return to water within the space of a few hours.

Then follows the desperate battle for survival as the hatchlings head unerringly for the sea, terrorised and often consumed by the many predators aerial and terrestrial which take a heavy toll of their numbers.

I saw only extracts of the film of the life of the Ridley turtle but I was enormously impressed and sobered by the vast unfamiliar world of sea inhabitants which David Hughes has brought into sharp focus in an era when these last remaining species of sea-turtle might so easily disappear.

Okavango

Having once visited the Okavango region of Botswana and revelled in the wealth of every form of wildlife it contains, I was doubly interested to hear that wildlife cameraman and zoologist Dr David Hughes had just returned from a filming and fact-finding expedition orientated towards gaining a true assessment of the threatened natural elements within the Okavango Delta.

A complete range of African game animals is attracted to the lush, fertile delta, end point of the Okavango River, one of the largest in southern Africa. It is unique in that it does not spill out into the sea, as do most other rivers, but empties itself into what would otherwise be arid bushveld.

David's film deals primarily with the problem of water usage for the mining industry, within an environment of fragile but perfect balance. Is this being done with caution and consideration so that the wealth of fauna and flora will not be unduly disturbed? And how about the Moremi Game Reserve which lies within the delta? Can one assume that the wildlife within its borders will remain secure from the effects of a disturbed water-flow?

David believes that disturbance of the water-flow is less of a problem than that of human infiltration, particularly now that

plans to control the tsetse fly are being set in motion.

'This will mean the end of the game and the invasion of man and his cattle into game-rich country.'

When insecticides are used for the eradication of the tsetse, side-effects cannot but have a toxic effect on fish and other aquatic organisms, and indeed, on the whole chain of life which radiates from the swamp area – including the herds of lechwe, the sitatunga, the bushbuck. But it is the richness of bird life of every description that would eventually suffer the effects of environmental disturbance. Those birds which depend on water-dwellers for their sustenance, such as the fishing birds, with techniques as varied and remarkable as their appearance, may never recover from human disturbance.

Each species practises a different method of food-gathering. The naked-head marabou stork, so often found in association with vultures, adapts from carcass-scavenging to insect-eating to fish-catching, using his powerful bill with deadly accuracy.

The red-legged spoonbill probes for water-insects in the shallow mud, sweeping and swinging its whole body from side to side as it moves along its rhythmic way.

Egrets and herons, the black variety, canopy their wings into a circle of shade which lures their prey within easy range. For a moment a small group stands on the alert, with folded wings, assessing their hunting terrain, judging their distance and their next point of attack. In the next instant they move low over the water, heads down, wings spread into a wide cloak, their well-proven method of predation as rare and beautiful as the humming, life-endowed haven of the Okavango swamps.

'If there's any drama in my life that I could wish to relate then this is it. A young otter playing in the water, a small snake taking termites from a mound, and those birds framed against the crystal-clear water – these surely are Nature's peerless wonders.'

Will David's message and that of other ecologists reach out and caution man against marring one of the last remaining outposts of nature?

I wonder how the Okavango story will unfold over the next ten years?

Spirit of the Helderberg

We were high up, though not by any means at the highest point of the Helderberg Nature Reserve, which tops even Table Mountain at 1,142 metres, viewing that breathtaking world of sheer cliff, wild flower and tree slopes and, below us, the vista of fathomless ocean and the silhouette of Cape Point.

'These mountains really get you,' said Douglas Wood, horticulturist Warden of the sanctuary, introducing us to his beloved land step by step.

'When you're up here, you feel really insignificant: you lose your stresses and your problems and see life in perspective.'

I knew only too well what he meant, for the bushveld wilderness produces the same ego-shrinking effect as did this terrain, its endless rhythm of the tides contrasting with the static fortress of the horseshoe range of the Hottentots-Holland basin which faced us. Having fought to tame this tangle of vegetation since 1961, so that, five years later, the Reserve became accessible to visitors, Douglas Wood now knows his sanctuary inch by inch, from the mountain crests to the deepest valleys.

Helderberg is many things and one could spend days discovering them. In three hours Douglas gave us a wonderful review of the history, the heartaches and the value of this relatively newly-protected area which the public is able to enjoy with him.

Wisely, Helderberg is administered by a democratic Advisory Board which includes members of the Somerset West public, but it is also controlled by the Nature Conservation Department of the Cape.

Driving up the steep, curving slope, not normally open to the public, we were enchanted to see a sleek herd of springbok

descending to one of the waterpans, soon joined by a smaller herd of eland and black wildebeest, each animal imported into the Reserve and carefully observed for successful adaptation over the years.

There are duiker, steenbok, caracal lynx, small spotted genet, dassies of the rock variety, which no doubt provide food for the black eagle nesting in the upper rockfaces and for leopard, whose spoor and claw-marks have been recorded many times.

Over 170 species of birds have been check-listed including kingfishers, woodpeckers, sunbirds and guinea-fowl, which nest among the Proteas.

We began our morning tour of the Reserve on foot, and although Douglas Wood assured us that February is an 'off' month for flora, the well-tended gardens near the entrance were an aromatic and visual joy. The scent of cloves and thyme emanated from the blooms of *Salvia africana* and from the nearby tall wild almond tree, a species originally planted by Van Riebeeck at Kirstenbosch. This specimen, we were told, is centuries old, watered by a perennial stream, one of many within Helderberg. The flowers all around us were spectacular: over two metres tall, the golden heads of the pin-cushion Protea (*Leucospermum*) shimmered in the soft sunlight but could not overshadow the velvet-soft limbs of the feathered shrub *Phylica pubescense*, which is endowed with the softest, green-toned texture of leaf I have ever seen; the now flowerless *Protea scolymocephala* glowed with red-tipped autumnal leaves and here, too, were the giant pink heads of *Protea cynaroides*, perhaps one of the loveliest flowers in Africa and one which blooms from midsummer to spring.

'Majestic flower to prompt the thoughts of men, according to their nature and their dreaming,' wrote South African botanist Sima Eliovson of the Protea.

How can one feel anything but lyrical at the sight of so much beauty?

I asked Douglas Wood about the plaque set at the archway entrance of the Reserve: 'Have time to dream, to think, to enjoy the peace and quiet and the wonders of life – the sea, the mountain, the fresh air'. But he could not divulge the name of the author, who wishes to remain anonymous. Whoever he is, he has certainly caught the spirit of the Helderberg.

Pygmy Retreat

Sitting in the mid-morning sunlight watching a flock of water-birds in the Jonkershoek Valley, one of the Nature Conservation Stations of the Cape Province, made me wonder, once again, whether our water-fowl and their wetlands environment would, eventually, be fully appreciated and protected. Not only in Africa but the world over.

I had recently been told that in the next two decades every major river in southern Africa will have been dammed, an asset to man but often a liability to water-fowl due to resultant habitat destruction. In view of this, the future of water-birds depends on their ability to exploit artificially-created wetlands. Are we doing enough to help the re-establishment of wetland refuges? Are we aware of our close ties with water-fowl over the ages, aesthetically, for food, for recreation? Do we take into account that water-fowl provide direct economic benefits since they eat the snail vectors of bilharzia, for instance, and grub up the corms of Cyperus, today's most important weed menace the world over.

The vivacious little flock of 15 pygmy (dwarf) geese skimmed and glided effortlessly across the water top, their colour-markings startling in contrast to the surrounding green-brown aquatic vegetation. Smallest of all water-fowl in the world, this un-goose-like goose bears brilliantly-distinctive decorations, the male displaying deep iridescence above, four separate head colours and chestnut brown along the neck and flanks. Strange to think of a goose nesting in tree holes or along cliffs, or in ant-hills, I reflected as we soaked up the morning sun that shimmered across the Jonkershoek mountain-side. Fossil records of water-fowl date back 80 million years. Had this little pygmy goose, now prized as very rare among conservationists, existed in those earliest times?

In this age of diminishing wild animal life it is good to know that these geese, recently imported from Botswana, are being treated as VIPs, diminutive though they are. Here, at the breeding station, they are completely safe, their extensive aqua-enclosure protected from aerial and water-level predators. James Millar, in charge of the station, explained the detailed plan of the breeding scheme. It is the future that must be considered, he said, a time when the cry of the Egyptian goose or the hunting tactics of the remarkable black heron might be no more than a few lines in a book.

Creating a central reservoir for species such as Cape teal, African pochard, the Maccoa duck, and Cape shoveler, has become a full-time research and management activity, for it involves new knowledge of capture techniques, captivity psychology, suitable feeding schemes, discovering a method of faithful and reliable day-to-day recording of behaviour traits, sexual cycles, lifespan and disease trends.

When adult birds are finally released back into the wild, individually marked for future observation, new problems and life-styles emerge. There are the undeniable effects of the marking rings on other water-fowl and their possible resentment towards the intruders. We hear a great deal about large wild animals released back into the wild but rarely do we get news of the lesser species, such as water-fowl, undergoing similar and just as vital rehabilitation programmes, and this in spite of the fact that the work involved is just as demanding of the biologist's attention and skill.

The week-old hatchling Maccoa duck which Caspar Heyl, in charge of that project, held in his hand for us to inspect was unbelievably fragile, and enormously appealing. Hatching successes now rate as high as 70 per cent, which is amazing when you consider that in the wild predation and other disasters probably take enormous toll of pre- and post-hatching life.

The Maccoa is very rare. The Jonkershoek Research Station project ensures its continued survival, its biological investigations ranging from the Cape vulture, kudu, Cape baboon, grey squirrel, the Cape laughing dove, to the rare, elusive Cape mountain marsh rose, as assiduously protected as those enchanting, vivacious dwarf geese.

Mercy Trap

If your livelihood does not depend on the raising of stock and crops, it's easy to be outraged at the thought of wild animals being killed, particularly when protected species such as caracal lynx and black-backed jackal are involved.

In a park or reserve stringent measures are adopted to safeguard all fauna; then why, many ask, is it permitted to hunt and destroy those very same creatures simply because they happen to reside in terrain which does not lie within the safety of a sanctuary?

Farmers are apt to get a little hot-headed on this issue, as does the public, especially when the term vermin is applied to such animals as cheetah or lion. Surely, nature lovers insist, the farmer can afford to lose a lamb or two or even more in the interest of the greater issue of conservation.

Is the public over-sentimental? Are farmers unnecessarily callous?

In the Western Cape the control of stock and crop losses is part of conservation practice and perhaps one of the Department of Nature Conservation's most difficult functions.

'It is surely in this field that a law-enforcing government body is tested to the utmost in terms of firm, yet gentle public relations service. It is the only real way to persuade and educate farmers into an ecologically aware, responsible way of thought.' So Dr Neil Farrell, in charge of research, explained to me as we drove to the 'problem animal' (the term vermin has been officially abolished) Nature Conservation Centre at Robertson, known as Vrolijkheid Station.

It was over two hours by car from my seafront base at Gordon's Bay. We passed through spectacular scenery reaching Villiersdorp just as clouds were gathering along the cliff-top crests and a gale-force wind began to blow from the south-east. The famous Elgin apple paradise – which that night was to lose 80 per cent of its crop in the storm – then came into view as we made our way to the shoreline of the Brandvlei Dam speckled with great numbers of Egyptian geese and guinea-fowl. Eventually we reached the Little Karoo flats flanked by the more distant formations of the Langeberge.

It seems that the Department's nature conservation policy covers a very wide range of responsibilities with a minimum number of officers and experts. Sven Phillips, management officer of 20 years' standing, proudly pointed out the many types of highly trained hunting hounds, highlighting the need for peak-performance packs when wild animal elimination was called for.

Since the need for control is ever-present, humane, controlled hunting is practised. The station provides funding, assistance, advice, on-the-spot guidance when this is needed, and keeps track of exactly which farmer is doing what within the Province.

Traps and other controls are tested, the results recorded for an overall census and possible alteration. Scent-bait traps (still under research) to prevent the destruction of non-harmful animals such as monitor lizards and mongooses will eventually be used. Crude traps, irresponsibly administered poisons, inhumane hunting methods are gradually being eliminated. Instead of unsympathetic coercive law enforcement, moderate,

courteous link-up methods of communication are beginning to pay dividends.

Sven Phillips has his own theory about the lynx's role as a stock killer, certain that two separate types of caracal lynx exist. He believes that the smaller, red-hued variety is guiltless of stock-theft, ranging less widely than its long-legged lighter-skinned counterpart.

It is obvious that neither Sven nor any other conservation officer takes pleasure in animal control, whether through hunting, or the cyanamide-loaded 'coyote-getter' that brings swift and certain death.

To them, it is all part of the job and their understanding of and interest in nature are probably just as great as that of the man who is fortunate enough to be taking part in restoring a segment of a wilderness.

Until I went there I wasn't certain, but now I am, and I believe, as B. R. Roberts of the University of the Orange Free State said, that 'the real conservationist is a special kind of person, with sound motivations, patient confidence and, above all, a compassionate understanding of and a humble, unselfish approach to his God-given environment.'

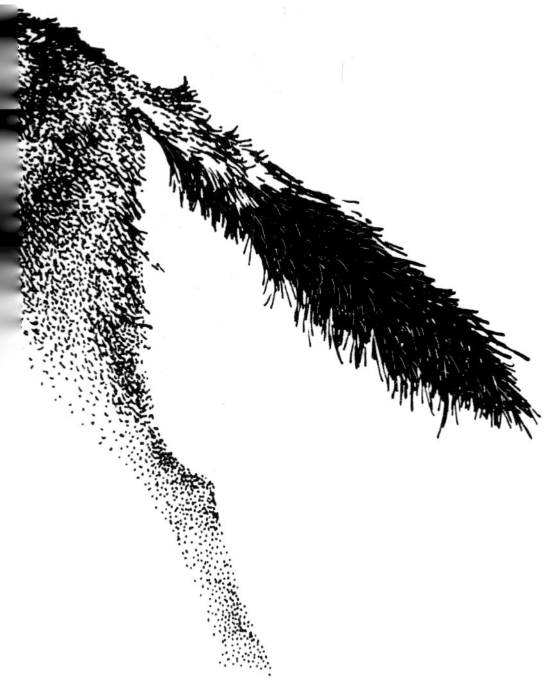

Marsh Rose of the Mountains

With a shyness characteristic of those whose work takes them far from the madding crowd, botanist Phillip van der Merwe, of the Assegaibosch Western Cape Nature Conservation Research Station, described the dilemma of the marsh rose (*Orothamnus zeyheri*), an almost extinct member of the Protea family.

'This unique flower,' said our host with sadness, 'really is on the danger list and we must face the fact that we may not be able to save it.'

Like the proverbial will-o'-the-wisp, this mist-addicted statuesque plant grows on the mountain slopes near Caledon. Ephemeral, elusive, here today and perhaps gone tomorrow, it is by nature restricted to one type of habitat where, since 1884, it has been discovered, lost and rediscovered.

Preferring peaty, spongy soil it is too easily spotted by the passer-by, for it may grow to a height of 2 or even 3 metres. And who could ignore the pomegranate-red flower, touched by a fringe-tip of soft white hairs, petals folded upon one another exactly like the petals of a true rose?

Imagine coming upon a cluster of these translucent, magical flower-heads after a steep, hot climb to the top of the forbidding, weather-chiselled slopes. Would you be content to sit and worship at their feet, or would you be tempted to pluck the long stem and carry it home?

I am certain that you would not, if you knew the secret of the marsh rose, for once the long stem has been snapped the life force of the entire mother plant is irrevocably destroyed.

Perhaps if it had not been for a man named Rodney Baxter, nature conservation officer of the Cape department, even the few remnants of the marsh rose would have been irretrievably lost.

Initiating the first intensive field study of the rose, he instigated the creation of an unofficially protected reserve, which at the present time contains 51 treasured plants within 20 hectares of wild land, concentrated into a close-knit and hence very vulnerable community.

Unfortunately, as so often happens, the care and strict protection of the marsh rose in the wild cannot alone ensure its survival. What is needed is a nucleus of plants, and a method of nurturing them from seed to adult.

Phillip van der Merwe's research project consists of both field and close-up nursery investigation in the form of grafting trials, using an inter-generically suitable stock-plant such as the 'pin-cushion' Protea (*Leucospermum conocarpodendron*).

The choice of a suitable parent plant is never easy for it involves detailed, microscopic study of the chromosome structure to determine harmonious growth factors (between stock and grafting), and root-top composition among other things. Quite apart from parent plant choice, there follows the vital problem of recreating a suitable climate. The presence of mist rates very high in marsh rose priorities, which means that density, periodic dispersion and rhythm of mist flow must be measured and constantly administered. Singeing heat or desiccating wind can spell doom to a sensitive rose at any stage of its life.

There is another side to the successful rearing of these rarities, one that requires more than scientific know-how or the ability to cope with tough mountain conditions all the year round.

'Do you believe that your plants respond to your own force of personality, your own special kind of care and attention and loving concern?'

The question of plant-life sensitivity used to be regarded as remote and even eccentric, but surely that time is passing, especially as proof of plant-to-human emanation and response is being accumulated in many parts of the world.

Botanist van der Merwe replied to my question very simply. His knowledge is based on practical results gained over the years.

'Five of my grafted plants "took" at first attempt. You can say that was almost a miracle. And yet, when I was away on vacation for three weeks and discontinued my weekly round of inspection of all plants, five died in my absence. Now I see them regularly and not one has died since.'

He would commit himself no further, allowing me to draw my own conclusions. There is no doubt in my mind that his devotion to those loveliest of flowers will keep them living.

Cape of Good Hope

How well T. V. Bulpin's thoughts sum up the initial impact I myself experienced when I first saw the Nature Reserve contained within the most celebrated cape in the world, the Cape of Good Hope:

'. . . at the south-west end of a great continent, the blending point of cultures of East and West, as well as of the currents of two great oceans, a remarkable richness of flora, fauna and a spacious, airy beauty of landscape combine in the Cape of Good Hope with a most tangible atmosphere of romance.'

How, I asked myself on that morning of windswept horizons, how can one adequately describe the immensity, the towering, age-eroded cliffs, the haunting, ghost-like silhouettes of sculptured stone that evoke images of dream-like fantasy? Here man must surely marvel at the elemental timelessness of the earth and feel overwhelmed by his own paltry insignificance as compared with the granite-studded sloping dunes endlessly challenged by the onslaught of the restless ocean.

Where else, in all of Africa but here, could you feel free and safe to walk unrestricted in the shadow of those coastal giants, with the scent of the sea breeze in your nostrils and the thudding hooves of sleek, regal eland and exuberant zebra in your ears?

Down there, as I looked among the fynbos vegetation so typical of this part of the Cape, a pair of ornate ostrich fed as they moved along apparently totally absorbed, yet probably aware of the high-pitched cries of the gulls, which compete with the beat of the surf.

Within sight of Kromriver I sat on a sand dune, looking toward the Atlantic, with the expanse of sanctuary stretching well out of sight. Not far from where I rested, baboons, not at all concerned at our presence, were feeding on overripe sour figs. Under my hand a small, deep red Euphorbia, delicately white-tipped, glistened with the vitality of health, gaining its nourishment in some inexplicable way from the barren sand.

Gerald Wright, Chief Warden of this Cape Divisional Councildirected Reserve, related the history of the area and of the sanctuary as we spent a head-spinning day of exploration, covering a very large section of the whole, driving sometimes along tarred roads and sometimes along rock-studded tracks of wild land.

In 1914 a powerful lighthouse had been installed 70 metres above the sea at Cape Point. Now its 19-million candlepower makes it the most powerful lighthouse in the world, sited where countless vessels, some still visible from the shores of the reserve, have come to grief over the centuries.

Not until 1938, however, did the nature reserve come into being, as the result of more than 10 years of struggle by a small band of enlightened men. Fencing, law enforcement, road-making and an overall management plan were gradually evolved as more and more land was added to the whole through the purchase of farms. Once described as a 'wasteland, waterless and treeless, with not enough to keep a scorpion alive', it was transformed into a wonderland for nature-hungry tourists, with access to shore points, picnic spots and a restaurant to serve day visitors.

During one December 67,000 visitors of all races visited the reserve, at a charge of 20 cents a day, which includes a road map and brochure giving information about bird life, plants, shipwrecks and wild animals.

USE YOUR EYES says the leaflet, for here, if you take the time to search, you may see black wildebeest, Vaal rhebuck, grysbok, baboons, porcupines, and myriads of tiny wild flowers. You can walk along tracks and footpaths, from Cape Point to the Cape of Good Hope, to the mouth of the lagoon known as Die Mond, along the Maclear Beach, littered with a variety of driftwood, flotsam and jetsam.

To the west, the sea is dotted with glistening islands of kelp. A fin, far out at sea, could be a whale skirting the land.

In this land of contrasts fragile, gentle Disa flowers bear the gale force as do the granite rocks of Cape Point, exposed to the full rape of the ocean currents.

According to legend Adamastor, one of the hundred giants who rebelled against the gods of ancient Greece, was condemned to be transmuted into these very rocks, his pagan spirit forever haunting his tomb which, so it is said, forms the peninsula of the Cape of Good Hope.

Tsitsikama

'The sea is the last of earth's undiscovered realms,' writes Dr Gilbert Robinson, the nature conservator of the Tsitsikama Forest and Coastal National Park. Turbulent, unfettered, vast, it is a wilderness in the fullest sense. Its shorelines and coastal belt are as untamed and magnificent as any southern African inland wilderness.

Here, where eroding rocks and shifting sands are battered by the ocean, the wonder of marine scenery may distract the visitor from taking a deeper look at the environment which contains more varied and perhaps even more beautiful fauna and flora than he is accustomed to seeing in inland game country.

As in the bushveld, the sounds, scents, textures and colour tones respond to the impact of wind, sun and rain, which dominate the life-pattern of every living thing, visible and invisible to the human eye. Within a few hours or less the mercurial ocean can change its identity: the shimmering, velvet smoothness of the surface may be stirred into an inky blue-black sway of twisting churning crests that may defeat the mariner yet does not dislocate the myriad life-forms which are part of the fathomless almost unexplored deep.

Floating on my back within the shelter of Mooibaai, I was very much aware of the thundering rollers' impact on the seaboard, contrasting with the gentle caress of ripples against rock-ledges on each side.

Where the incoming tide gradually encroached upon the beach, the hiss of toppling waves and sucking undercurrents met the softness of the sand like the weathered palms of gnarled

hands rubbing against one another. Facing away from the shore, I followed the flight of an eagle, perhaps the local fish eagle, and a pair of Cape cormorants diving for fish.

Many trails have been blazed where steep cliffs rise up to the forest, so that the visitor to the park can savour the grandeur of the yellow-woods, red currant and elder, Cape beech, chestnut – 27 trees and shrub types numbered for identification along the paths.

Suddenly the mid-morning sun was dimmed by a blanket of cloud and I became aware of the chill of the water and the eyes of a long-tailed grey-brown mongoose upon me from the distant slope of the beach. I wondered if I would see the clawless otter, whose footprints we had seen the previous day; delicate, finger-like impressions in the sand.

Dr Robinson had led us along a section of the four-day otter trail which takes the hiker from Storms River to Groot River, crossing ravines, winding through indigenous forests, fording numerous streams. Hugging the Tsitsikama coast as closely as possible, this trail is the answer to any nature-lover's dream – routed through virgin forest, exploring the wonder of tidal pools up and down inclines of glowing Ericas, Proteas, Knysna lilies, all within earshot of the breakers and the constant chorus of birds.

In spite of late afternoon turbulence, Dr Robinson had steered our boat safely into the calmer waters of the Storms River mouth which flows between the rock-layered cliff face of Table Mountain sandstone.

Within a cave his powerful light momentarily illuminated the eyes of hundreds of clustering, upside-down clinging bats of three types, the focus of extensive research under unusually hazardous conditions.

The fish eagle pair had remained quite unperturbed at the sound of our boat engine within their gorge. 20 metres deep at the narrowest point, the echo-sounding device traced the exact river depth as we moved back towards the sea.

Tsitsikama, South Africa's first Coastal National Park – they say it means clear water, or whispering raindrops in the forest – is a sanctuary of incredible beauty which should be explored down to the last magic tidal pool, always with the guidance of the many available books, or better still, in the company of an experienced ranger.

Dawn had just broken when we began our exploration of the inter-tidal zone which reaches its upper level at the spring tide. On every side of us and underfoot, as we crunched over thousands of welded mussel colonies and honeycombed beds of tubeworms, there was a profusion of life.

After a time our unseasoned eyes became attuned to the distinct bands of animal and plant forms along the shore, their very existence totally dependent upon the delicately balanced environment geared to temperature, solar radiation, food supply, the nature of the sea bottom, and the quality and rhythm of the tides, alternatively exposing and covering the marine inhabitants.

The distant infra-tidal zone, set against the curling breakers, exposed shapes and forms of vivid colours, from velvet black to red, to crimson, to the yellow-gold of sea-slugs. There, too, we found the blue-bottle (Portuguese man-of-war) so transparent-blue that its tint renders it invisible to predators. The blue-bottle is not a single animal, but a colony of individuals, each one modified to fulfil a special function such as reproduction, feeding and protection by paralysing its prey. The toxin injected by the stinging cells of the colony individuals is similar to cobra venom and 75 per cent as potent, though your life is not in danger if you are injected from a single sting.

The sea-anemone too (you might well ask whether it is plant or animal) belongs to the divers group of *Coelenterata*, which includes jelly-fish and fern-like hydroids. Tinted red, green, pale translucent cream, the anemone's delicate, flower-like appearance and gold tentacles mask its carnivorous habits. Paralysing its prey with stinging cells contained within its slender fingers it can, in an instant, recoil into a homogeneous lump if threatened by the incoming tide.

Through Dr Robinson's expert eyes we beheld myriad shapes of seaweed and sponges, and spiny starfish forms, startlingly bright and easily visible in their five-fold symmetry, each segment being completely independent, and able, in the event of injury, to regenerate any part of the whole.

We were shown cucumbers, urchins, an octopus lurking in a

submerged cavern, invisible to all but the most skilled observer.

'Listen to those rocks!'

Diana Calburn, wildlife feature writer and my travelling companion, was standing stock-still in sudden awareness of the repetitive, crepitating, clicking crackling of the vast barnacle colony adjusting for the millionth time to the recession of the tide and its consequent exposure to air.

Dense masses of fawn to light-brown crustaceans, whose life-cycle includes a free-swimming period in the plankton before settling down on the rock-face, were securely cemented and protected by a shell of plates which responds day in day out to the touch of the tide.

Here too, were Venus Ears, cockles, scallops, flat-worms and the largest of the chitons known as the armadillo, its plates of shell surrounded by a girdle of immense firmness, as you would soon find if you attempted to prise it from its rock grip.

Within the sanctuary, nothing may be touched or moved. Should curiosity, or your admiration, perhaps, for a rock-crab or a glinting stone decorated by russet lichen, overcome you, remember that many marine plants such as algae grow on top because they must absorb radiation from the sun. If you do not replace the rock the way you found it (as Dr Robinson explained) you may expose the creatures sheltering beneath it, and at the same time destroy the algae growing on top, thus depriving yet other animals of part of their food supply.

For days after our rock-trail among the tidal pools, the impact of those hours persisted in so many different ways: sudden muscle aches in unfamiliar places; cheeks glowing from the sting of the sea breeze; the sight of the early morning sky, flushed with new-day pink; the smell of the ocean, merging with the aroma of forest; the swift, decisive swish of the tidal ebb and flow which determines the life-pattern within the inter-tidal zone and in the tidal pools.

If you have a few days, if you can take your time searching the low-tide expanses at your leisure, you will come away with an overwhelming sense of the grandeur and diversity and wonder of the ocean shore.

The Addo Tuskers

When, in 1919, the famous hunter Major Pretorious was called in to eradicate the last of the elephants which threatened the agricultural progress of the Sundays River Valley in the Eastern Cape, the fate of those Addo giants seemed irrevocably sealed. And yet, due to widely aroused human compassion and a handful of enlightened men who at the eleventh hour called a halt to the slaughter, the Addo National Park fence now securely encloses the herd in a terrain vividly described by the great hunter himself, who succeeded in killing 120 elephants in eleven months: '. . . I visited Addo and soon realised that if there was a hunter's hell, here it was – a hundred square miles or so of all you would think bad in Central Africa, lifted up as if by some Titan and plonked down in the Cape Province. It was scrub, generally some eighteen feet high and exceedingly thick. Once in this jungle it was seldom possible to see more than five paces ahead . . . A terrible country!'

Driven from their ancient, vast homeland into 10 remaining square miles of bush, the elephants had become dangerous and wily in the extreme, for the time being defeating even Major Pretorious. When only 15 tuskers remained the sympathy of the public was suddenly aroused for the heroic hunted band which was to undergo many trials and tribulations before Harold Trollope of the Parks Board eventually succeeded in driving the herd into the area proclaimed as the Addo National Park in 1931.

After my recent travels in the Western Cape of South Africa, thick cover like this, situated barely 70 miles from the thriving coastal city of Port Elizabeth, reaching from the Suurberg Mountains to the Sundays River Valley of agricultural enterprise, evoked a new kind of wilderness atmosphere. Here the evergreen vegetation is a total contrast to the 'red' elephants tinted by the soil. How different from a typical bush scene in my own home-base Eastern Transvaal lowveld, where the grey herds of tuskers are easily lost among gnarled stems and canopied tree-crowns.

It was late afternoon when we arrived at the gate of Addo

National Park, a time when the chance of an animal encounter was very slim. We made our way to the Caesar's Dam road which passes a look-out hillock and at once the vastness of the bush country opened up before us.

We saw no game yet we knew they were all in there: birds, reptiles, mammals, great and small, and above all, the elephant herd of 74, the preservation of which has become an epic in the history of game conservation.

Here the forest canopy, up to 3.6 metres in height, 6,852 hectares in extent, possesses a quality of lushness I have never seen before, not even in East Africa. Here you will find *Portulacaria afra*, known as spekboom, the favourite food of the elephants, with mimosa and succulents of many types, Stapelia, wild pomegranate, Crassula and Mesembryanthemum species; here are the crimson blooms of tree fuchsia, sneezewood and flaming red flashes of the torch lily, whose visual impact is created by masses of closely-packed filaments, not unlike the wisp of a paintbrush, growing straight out of the ground without a hint of green leaf.

Many types of habitat are found at Addo: glades of open grassland for buffalo, hartebeest and other grazers; low bush for browsers such as kudu and black rhinoceros; plenty of water-points attracting a wealth of bird life and nocturnal residents such as antbears, caracals and bushpigs.

Rounding a curve we were startled and utterly delighted to discover one lone tusker so close to the cable and tram-line fence that it seemed we could have stretched out our hands to touch his pinkish rust-coloured hide. Where in all Africa (except in a zoo) could one ever venture so close in complete safety, removed from that special realm by nothing more than a barrier of metal and wood?

Amazing to think back to those years, from 1934 to 1953, when one after another restraining fence had been erected and discarded as useless in the face of elephant ingenuity, until eventually the rail and cable tension construction had defeated any further efforts of elephantine destruction.

Contentedly we settled into our car seats and revelled in the spectacle before us, unperturbed by the knowledge that the young male, short straight tusked and mischievously arrogant, was very much aware of our presence, which he was graciously prepared to tolerate provided we remained passive observers and did not disturb the peace.

He had moved into a grass patch for the next course of his endless repast, curling his agile trunk around each tuftful and with utmost skill and care testing, plucking, scratching, blowing, scenting. His marvellous appendage could do all these things and more; in an instant, when need arises, it becomes transformed into as lethal a weapon as a crocodile's tail.

Not far away, to the south-west, the most southerly remnant of Africa's elephants struggle vainly for survival under the increasing pressures of human encroachment. The Knysna elephants are now, in a sense, where their Addo relatives were before Park Ranger Harold Trollope drove them to the safety of permanent refuge. For some time the plight of these benign Knysna mammoths has been stop-press news within wildlife circles, for which reason, some years back, ex-ranger author Nick Carter undertook to conduct a survey of their terrain, social structure, food habits and movement pattern.

Sponsored by the Wild Life Protection Society of Southern Africa, his year-long investigation led to certain resolutions and recommendations being adopted but they yet remain to be implemented. They should be, for as time passes the existence of these heroic survivors hangs by the slenderest of threads.

Ten remain, and of these only two are cows, a situation not unlike the crisis which evoked a complete turn-about in policy in favour of the Addo group. 'Overnight the sympathy for the hunter changed,' wrote a spokesman for the National Parks Board, 'and became concentrated on the hunted. Voices went up even from the outside world to save and conserve the once mighty herd which later started the Addo Reserve. *Viva* Addo!'

And why not *Viva* Knysna? *Viva* freedom and permanent safety for the handful of persecuted tuskers. Could they, like their dusty-pink Addo brethren up the coast, not be accorded a sanctuary of peace, there to breed, feed and wander for the rest of their days?

Among the Hunters

In a recent wildlife film of enormous public appeal, the hyena was once again presented as a diabolical, fiendish villain of darkness for no other reason than the traditional call for such a character, truly amazing in an era when research work has revealed quite another side to the spotted prowler of the bush. Greatly strengthened in my resolve to take up the cudgel in the cause of hyena justice by a recent short visit to the biologist Simon Bearder in his bush retreat, I decided to return in order to gather more up-to-date information that might put this legendary scavenger-predator in clearer perspective, for once and for all.

What Simon had to tell on the subject of his exploration into every aspect of *Crocuta crocuta*'s (the spotted hyena) life is astonishing, for he had uncovered a social pattern which reveals traits of the hyena's character which are very different from those usually associated with this elusive animal.

'Yes,' he affirmed in answer to my query. 'They *are* very intelligent. They possess the ability to assess complexities which may arise as a result of a social situation.'

His study had covered a very wide range, including the total jigsaw pattern of the hyena sharing his environment with jackal, lion, cheetah, antelope, porcupine, antbear and many others.

Driving out into the sanctuary with Simon and his Shangaan assistant, Abil, we checked at frequent intervals at already established sites for spoor (every item meticulously marked off on the prepared check-list) and for communal latrines. They aren't so hard to find as you might imagine because of the chalky-white nature of mineral and hairtrace droppings (the hyena's digestive system is particularly well designed for optimum food-absorption, the hair rejects being useful for prey-content analysis) and the 'scrapes' made by scratching with the front feet.

Interdigital and anal gland secretions provide another rendezvous information signal, perhaps as vital as vocal communication, essential in the species' finely tuned link-up routine.

Consistently correlating the evidence gained through day-by-day observation over a long period of time, in conjunction with his own highly developed recognition technique of individual members of two particular clans within his study terrain, Simon Bearder has built up a file of relevant data not necessarily applicable to other parts of Africa.

Within the hyena matriarchal society the young are carried in-utero from between 97 and 132 days, then suckled for nine months. The life-style is based on considerate, highly sociable and communicative behaviour, which must be fascinating to study and compare.

The sound system alone is absorbing and highly evolved but could well lead to the erroneous assumption, as Simon Bearder explained, that great numbers of the species exist when, in fact, only a few are vocalising with immense variation and force of sound, sometimes making it seem as though the whole forest were alive with them.

In one test area where only twelve individuals inhabit 30 square km, their territorial instincts and aggressiveness (when intruders infringe the scent- and sound-marked boundary) are violent and often as lethal as in other species, such as lion, in a similar situation.

'So often,' said Simon, 'the hyena is portrayed as cowardly. I wonder how that could have arisen?'

He told us of the time he and his wife were doing a night watch in mid-winter, as happens almost each day of the week except in the season of heavy rains.

At 1.35 a.m., after hours of expectation, they observed an adult leopard stalking a steenbuck not too far from a buffalo carcass on which two hyenas were feeding. Simon had been recording the pair at the kill, the sounds finally becoming non-aggressive: giggles, growls, barks, rumbles of greeting.

So stealthily had the leopard approached the unwary hyenas at their feast, that they, taken completely by surprise, gave hot chase to the intruder who only returned to what remained of the kill long after the hyenas had left.

And here is another tale, this time from the plains of East Africa's Serengeti, told by the famous biologist, George Schaller, in his book *The Serengeti Lion* (University of Chicago Press, 1972).

'Two lionesses have an adult wildebeest which they eat off and on, watched by only one hyena, then by several as more and more arrive silently.

At 2300, both lionesses suddenly begin to feed rapidly, as if anticipating the loss of their kill.

Within one minute 17 hyenas begin to whoop as if on signal and move in, circling the kill and drawing ever closer, until at 2315 their tails whip up and they rush at the lions. These trot off, emitting harsh growls as several hyenas chase them for 30 metres.'

Let's give the hyenas their due for courage and clan discipline and for knowing exactly how far they can go, with whom, when the moment is right – all this must be determined by many factors still to be discovered.

How is it that, at times, when hyenas have made the kill the king of beasts feels confident enough to contest their right and forces the hunters to give way to his dominance? Is this due to the season, the drive of hunger, individual mood, uncanny ability to assess the other's strength, or a mixture of all of these?

Shall we ever know even some of the answers?

Manyeleti

If ever you have had the good fortune to enjoy a clear, starlit bushveld night, the kind of night when the constellations seem to be within a handstretch, then you'll understand why Manyeleti has been so aptly named, for in Shangaan it means Place of the Stars. Sited west on the boundary of the Kruger National Park in the Eastern Transvaal lowveld, Manyeleti is a game reserve with a difference, catering only for African visitors.

It is staffed by an African team of 70, with the exception of the dynamic, totally dedicated Manager-Warden, Steve Kruger, whose long years in game country (he is an ex-Kenya District Commissioner) have endowed him with exactly the kind of

experience and understanding this post demands.

Lack of human disturbance, a profusion of every kind of game indigenous to the area, a first-class road system, a series of dams, successful counter-soil erosion measures providing excellent conditions for wildlife at every level from sub-soil to aerial, and a great spirit of newness and widening horizons make Manyeleti a refuge without parallel. It is hoped that under the guidance and direction of the initiating Bantu Administration Department similar reserves will be created in other parts of the country, for there is no doubt that wilderness experience must be provided for all sections of the community.

In eight years Manyeleti has become a thriving, popular reserve. Its camp can accommodate 540 visitors at one time catering for luxury accommodation as well as camping, at very much lower prices than those paid in other national parks. What a

wonderful opportunity it was for the many children of all ages I saw on the day of my visit, and for those adults who had never seen wild animals in their natural habitat.

Even to the non-urban visitor, much of what he sees in the sanctuary takes on a totally new meaning: the elephant is no longer what it has always been – a destroyer of crops; the leopard no longer a stock-thief; the lion a man-killer; the owl an omen of disaster.

From earliest years the White child grows up with picture books and toys of the animal world. The African child, though more closely linked with the bush, has probably never before had the opportunity of learning to appreciate the meaning of wilderness protection.

He can discover it here and in a few other centres within southern Africa, sponsored by enlightened bodies such as the South African Nature Foundation, the South African Wildlife Society, a band of private individuals, and of course, the Government's wing of Bantu Affairs.

What is unfolding at Manyeleti is at least as important as any phase of conservation or restoration within southern Africa.

Gentle guidance and a new look into the heart of the bushveld, this must be our greatest insurance against the destruction of our wild heritage; and this is only the beginning, for progress is slow but certain. Take the as yet bare-shelved museum, for instance, which in time will serve as an initiation to the wilderness, providing the visitor with the sight, sound and relevant information concerning the animals and birds he is about to observe living free.

Here he will see history and also pre-history, relating to a time when animals looked very different.

Across the way is the library to which he can turn in leisure hours to review and re-assess the experiences of the day.

At his side, kindly and helpful, Mr Nehemia Sithole, the reserve's liaison officer, is ready to answer any questions. His job is never-ending, for many would-be game addicts are starting from scratch and hardly know how to conduct themselves or how to go about getting the best out of their few days in the wilds.

Camouflage Enigma

Perhaps until that sunlit, sparkling day at Western Cape's De Hoop's Nature Reserve I had not fully understood the subtle effect or the wonder of camouflage. Dr Douglas Hey, director of the Cape Province's Department of Nature and Environmental Conservation, and one of South Africa's most highly acclaimed spokesmen for wildlife, had allowed me to accompany him on a day trip to that very special coastal belt of hauntingly beautiful country. De Hoop's 800 morgen are composed of wind-blown, vegetation-clad sand dunes which seem, at first glance, more like a series of undulating hills set against the shoreline of the Indian Ocean.

Here many different types of ecosystem converge. Here, also, within the white-walled green-lawned enclosure of the old farmstead, dating back to 1730, time seems to have stood still, though the house has been tastefully restored for the manager's use as a home, retaining the atmosphere of the early Cape.

This is the land of the wind and weather resistant milkwood tree, protected and beloved by birds and humans alike for the shelter and shade its thick foliage provides.

At the foot of the almost desiccated vlei-course we watched flocks of yellow-bill and Cape shoveler, Egyptian geese and pochards, alarming a pair of Vaal rhebuck that veered off between rustling stands of thatching grass.

In this area, bounded by the sea to the south and a ridge of limestone hills to the north, one could spend many days exploring the fynbos, the dune-veld, the calcareous rock formations and the limestone once submerged under the sea. But for me the highlight of that day was our early afternoon sighting of mountain zebra (*Equus zebra*), thought to have been indigenous here as well as inland in the mountains of the Great Karoo.

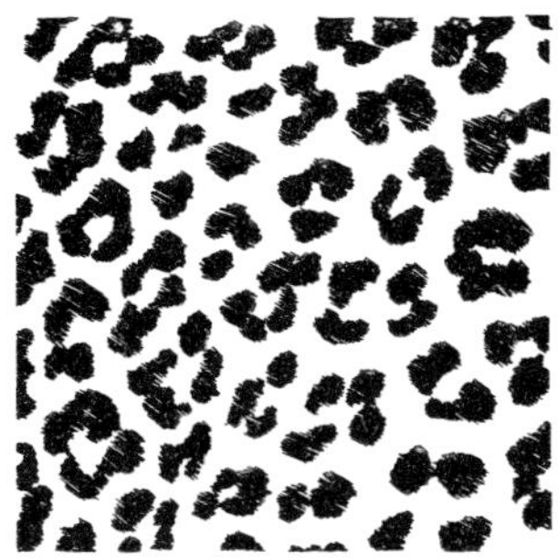

They were a long distance away, clearly visible only through binoculars, yet my first view of this almost extinct animal was at least as thrilling as coming upon a herd of elephants or a group of cheetah. They stood very still against the cleft of the dune-rise, as if aware of our scrutiny, facing us with their long ears raised, cupped forward and alert, in spite of the absence of predators in the Reserve.

After some little time, as one of the pair trotted off over the brow of the hill, a slow-moving cloud shadowed the scene, altering the colour and light textures within a matter of moments. Where there had been stripes and long ears and flicking tails, nothing but hillside vegetation remained, as if the zebra had literally been swallowed up by the solid earth.

What brought about this perfection of concealing coloration?

Hugh Cott of the University of Cambridge, a world authority on animal camouflage, has collated evidence to show that colour variation in mammalian species is almost certainly due to the elimination by predators of individuals that do not match their background as effectively as others. Thus, in desert regions, many small mammals are a much lighter and sandier colour than their close cousins, who may live in a darker environment.

Writers have commented on the way in which the zebra's stripes and the giraffe's blotches make these animals almost invisible in the thin cover where they habitually live, especially at dusk when an attack is most likely to occur.

Rudyard Kipling, in his famous story 'How the Leopard got his Spots', gives a delightful rendering of the whole system of natural camouflage. To avoid the leopard, the giraffe and zebra retreated into the forest and acquired blotches and stripes from the broken shadows caused by the sun shining through the leaves. By the time the leopard reached them they were so perfectly concealed that, in order to hunt them on equal terms, the leopard himself was forced to acquire a patterned coat of spotted fur.

Zebra of the Karoo

'They are part of history,' wrote Eve Palmer in her enchanting book *The Plains of Camdeboo* (Collins, 1966) in which she tells of her beloved Karoo, one of the world's strangest, most fascinating, yet most unfamiliar regions.

'They were the first zebras the Dutch ever saw at the Cape. By the middle of the 18th century the colonists knew them well, for wilderness was in their blood and they found that trying to tame a mountain zebra was a very different thing from domesticating a quagga. The zebra of the Karoo has escaped, but only just, the fate of the quagga.'

And now, within the space of a few hours, perhaps, I was to see them at close quarters, within the national park that had been especially created for their continued preservation.

Though I had moments of nostalgia for the gaunt, towering cliffs, for the rush of surf battering my body, for another glimpse of Tsitsikama National Park and for yet another dawn among the tidal pools, I felt a tremendous sense of excitement as the thick Addo bush receded below me.

Up and up, inland and north, I climbed the pass over the Suurberg, which follows exactly the original path of the great tuskers that are no more, those masterminds of speedy, economic, overland engineering. 27 km from Cradock I entered the gate of the Mountain Zebra National Park, situated on the northern slopes of the Bankberg Range.

For many years the fate of the zebras had hung in precarious balance, until in 1971 a heavy snowfall had tragically depleted the gradually increasing herd, numbering 11 in 1931, reaching a total of 25 by 1954.

I might have seen only a few of the family units of the mountain zebra in the short time I was able to spend in the sanctuary, but it was marvellous to know that somewhere within that 6,633-hectare refuge 158 zebras ranged over the valleys and hills without a predator to harass them.

Within half an hour of my arrival at the Berghof rest-camp – I

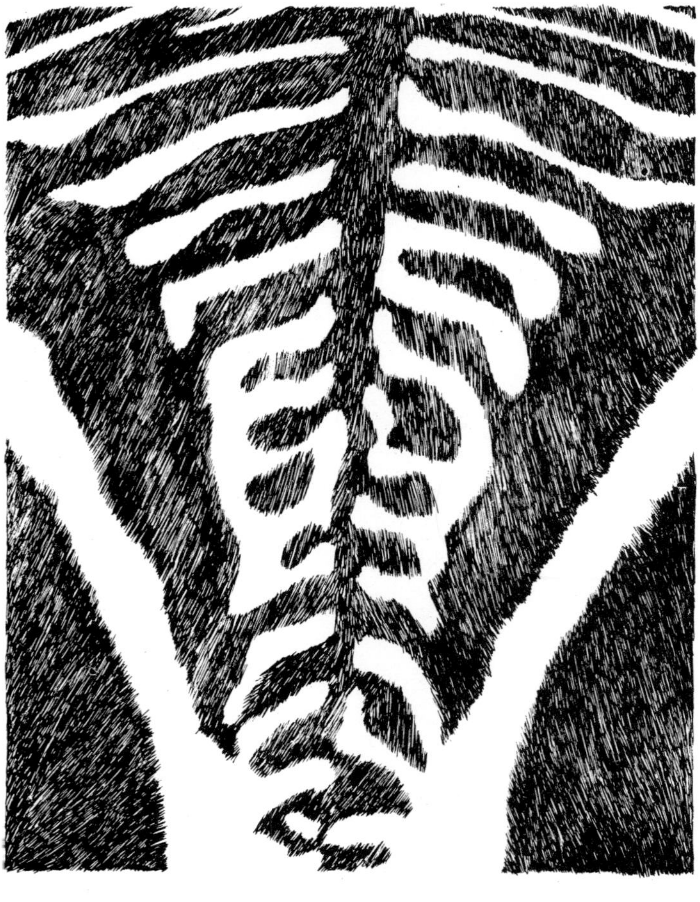

was the only visitor on that day – I found myself clambering on to the nearest hilltop an hour before sunset, enthralled by the overpowering tranquillity of the place. At each step another, yet higher, silhouette of mountain contour came into focus and with it a new play of shadows at the fringe of the sunset spectrum of the Great Karoo.

I remember being told about the utter remoteness, the other-planetary quality of this terrain. Below, towards the meadow-land near the resthouse, the world was already shrouded in the long shadows of evening. A klipspringer startled me out of my reverie, and I decided to descend from the land of the dassies, of tilting rocks, of sweet thorns and aloes and cabbage trees.

On the morrow Chief Conservator Piet Bronkhorst would be taking me to the feeding grounds of the herds.

The sunset hour I had spent alone in those glorious hills had given me an opportunity to appreciate the newness and the vastness of the country. It takes a little time for the eye to become accustomed to any new environment and especially in an area where the fauna merges perfectly with the back-drop as it does in Mountain Zebra Park. Here the hooved animals seem to blur into the grey and dun-coloured bushes and hillsides.

Driving up to the look-out post, a special feature of the park, Piet Bronkhorst explained the difference in zebra characteristics, for it isn't just a matter of stripes which distinguishes one type of zebra from another. I had grown used to seeing the Burchell variety. Here zebras bear a dewlap, longer asinine ears, black-brown muzzles, and stripes which are distinctly different in arrangement and do not cover the belly area.

Even the call of each zebra type has a sound of its own: the mountain zebra's call is more like that of a horse, and quite unlike the musical bark of the plains and bush zebras we know so well in other parts of southern Africa.

What a marvellous study this must have been for National Parks biologist Barney Penzhorn who, after years of intensive field study, was able to identify almost one hundred individual mountain zebras through a photographic identity card system. To know their living requirements, their social associations, their breeding cycles, is as vitally important for their continued preservation as is their actual physical protection and maintenance of their free-range environment. Without scientific knowledge to back up management procedures, long-term views or progressive policies could not be put into practice effectively.

They looked truly magnificent: groups of seven or eight, usually including one dominant stallion, juvenile, playful foals of different generations, and watchful, supple-limbed females, some well in foal.

Not far from them a gemsbok lay in the shade of a tree, its flashing white blaze glinting in the sunlight. Eland, red hartebeest, springbok moved past at a leisurely pace; two black eagles soared over the clifftops with astounding grace and speed.

Black wildebeest pranced as they marked time with impatient hooves, their pale-cream luxurious switches flowing to ground-level behind them, their movements deliberate, pugnacious, their 'extraordinary antics and capers', to quote Colonel Stevenson-Hamilton, 'forming a constant theme for anecdote among the early pioneers . . .'

Down in the valley, at the edge of a waterpan, a pair of blue cranes grazed alongside a herd of zebras. Just above them a troop of baboons loped their zigzag path from the lip of the hill, straight into a school of mongooses, among them a large entourage of infants scampering in all directions in the wake of their elders.

On the way back to camp that evening I met a zebra stallion face to face on the winding road, his quivering nostrils challenging my intrusion, tossing his head towards his herd as if commanding them to await his pleasure. Thus we confronted each other, until I switched on my engine and began to reverse, hoping that I might continue observing the group from a distance without disturbing them further. At that instant the great stallion snorted and leapt away from the gravel path, his curiosity and patience apparently exhausted. Over the next rise and down the slope they cantered, stopped once, turned back as if at a signal, then disappeared out of sight.

For a few more moments I thought I could hear the impact of their hooves. When I reached the top of the next hill and scanned the valley with my binoculars, they had completely disappeared from sight, enveloped by the shapes and colours and vegetation.

Kingdom of the Eagles

But for conservationists like Michael Jarvis of the Cape Department of Nature and Environmental Conservation, the fate of the great birds of prey might well be sealed within a few decades.

Have you noticed how very few they are these days, compared to, say, fifteen years ago? Yet the eagles and vultures are so much part of the wilderness, although few people realise their value or the remarkable cycle of their lives.

What is it that you feel as you lie on your back in the grass, your eyes intent on the graceful flight-trail of a crowned eagle soaring majestically without conscious effort, unhampered by such mundane limitations as gravity, or the need to refuel, or the observation of territorial boundaries which do not apply in his no-man's-land of thermal currents? Do you feel an intense desire to soar with him into his untamed world of space and freedom, no longer bound to the planet earth? Do you wonder that he's up there at all, in a time when man's unbelievable shortsightedness is taking dreadful toll of the great flightbirds of Africa? The farmers believe them to be destroyers of stock, yet research is proving that they are not, and this knowledge is currently being communicated to the farming community with some measure of success.

This is a long-term project, and a very absorbing one. The men who are carrying it out are gaining knowledge of the birds of prey as they go, for without precise, reliable information of such factors as feeding habits, the farming community is not likely to listen to the voices of the scientists. Reliable information means proof put across through good public relations and the ability to interpret the findings in the most palatable and practical manner.

I was astounded by the results of Mike Jarvis's fact-finding forays among some of the eagles, such as the black eagle, the martial eagle, and the crowned eagle. It took him over one year to complete just this part of his task, that is, visit nest sites, identify their occupants, collect bones from under the eagles' nests, finally proving how predominantly these raptors do feed on natural wild game animals.

Take number two of his study eagles of the crowned variety (his analysis table contains 14 detailed lists). This one had fed on: 9 hyrax (rock), 1 mongoose, 2 monkeys, 1 hare, 1 leguaan, 1 bird, 4 buck, 2 lamb kids, a total of 21 items, though it is not known over what period of time. I was fascinated to be permitted to examine several of these bone-bags, often identified from a tusk, as in the cane-rat, or a lower jaw bone, as from a young herbivore. Endless hours of examination, notation, correlation are needed to convince the farmers that the birds they are destroying are a vital segment of the whole ecosystem, maintaining natural equilibrium. They are almost entirely innocent of stock-theft and can easily be spared in the light of their overall benefit to Nature.

Mike Jarvis has also specialised in the study of the Cape Vulture, whose range extends farther south than any other African vulture, to Bredasdorp in the southern Cape Province. Flying out as far as 225 km from his colony each day, the vulture is never a predatory killer, as some seem to believe, but reduces the number of flies and fly-borne diseases of stock through effective scavenging of diseased, dying and dead animals. This is of vital importance in extensive farm areas such as the Karoo, where carcasses may go for days without being detected by man.

Perhaps of all large birds the vulture presents the epitome of a creature of contrast, graceful in the air, clumsy on the ground. 'When seen on Mother Earth, it is hard to realise', wrote Colonel Stevenson-Hamilton in *Wildlife in South Africa*, 'that this ungainly, clumsily hopping, repellent-looking bird is the same that so delighted our senses when on the wing, or sweeping straight down in a long inclined plane when his amazing keen sight indicates to him the presence of a dead animal.'

Mlilwane - Touch of Fairyland

Revisiting that gem of wildlife refuges in southern Africa, Mlilwane of Swaziland, was like coming home after a long journey and rediscovering well-remembered sights and scents and sounds which evoked memories of years ago.

A great deal had changed within the sanctuary, which has grown to ten times its original size and has been transformed into a profit-making, thriving animal reserve in a relatively short time. But the essence was the same – that intimate, special flavour of Mlilwane where each creature is regarded as a VIP, be it a rhinoceros, a giraffe or white-faced duck.

It was good to see such old familiars as matriarch warthog Lady Jane, now stiffer and older, but just the same old bristly character, using the camp as her own back garden where visitors are tolerated providing they respect her status as the oldest warthog resident.

I recognised Twinkletoes, once a ferocious, arrogant ostrich cock now transformed into a gentler, more sociable bird through years of marital bliss and the responsibilities of paternal care. There was the same bull hippopotamus, lonely as before, leading his bachelor existence among the raucous water-fowl, his temper amazingly pleasant considering his extreme sexual frustration.

Mike Morell – Warden Terry Reilly's assistant – took me on a dawn Land-Rover climb to the breathtakingly beautiful high point known as Poacher's View, from which, not so long ago, Terry dreamed his dream of new and larger vistas for his sanctuary. Metal-ingrained surfaces of granite below our feet told of the time when poachers sharpened their knives and spears in readiness for yet another foray against the already declining wild animal populations which eventually disappeared altogether.

Mlilwane Game Reserve dates back to 1962 when Terry Reilly set himself the enormous task of re-establishing a wilderness region, facing the challenge of re-introducing, acclimatising and managing a great number of imported species. Though these had once existed in that area, the newcomers had to re-adapt to conditions not nearly as favourable as their ancient, natural environment had provided. It was marvellous to see them successfully established and so unafraid in the presence of humans.

An impala ram, magnificent, alert, was preparing for the time of territorial marking, of exerting his male dominance, of rutting, of protecting his harem from the camp-following younger rams, always waiting for their chance with some specially appealing doe.

In the upper regions two giraffe shared the company of waterbuck, far removed from water, an unusual sight. Below road level stood a dusky-brown nyala bull, his snow-white chevron startling against the bush. What more beautiful or more aristocratic antelope is there to be found in Africa?

Nearby a blesbok bull was driving his three ladies before him with intense zest: tail up, neck stretched forward, he pranced and swaggered, shaking his head, dilating his nostrils, obviously feeling the primeval urge for procreation. A herd of zebra with young wandered nearby; along the Mhlambanyatsi River course

I passed a domed, many-layered hammerkop nest, very like a Swazi hut, neat and immaculate, its one-way entrance set at the base.

Perhaps the most enchanting creature I met was an orphan bushpig, not more than three weeks old, adopted into the elastic Reilly household of four small children plus a family of six hyrax.

Have you ever had your toes lovingly chewed by the short-tusked jaws of such a porcine infant, an experience so very different from the nudging bleating insistence of an impala calf or the overwhelming purring caress of a baby cheetah?

Walking on delicate tip-toe hooves, this foundling of the questing snout, passionate dark brown eyes, and pointed ears delighted us with his antics and his total devotion to the task of destroying somebody's pair of shoes.

Try to visit Mlilwane if you haven't done so already, especially if you are in need of enchantment and total peace. You'll see no lions or leopard or elephant, but there's so much else here to delight you that, like me, you will probably never want to leave. Of all the dreams come true, this one really has a touch of fairy-land about it.

Hippo Interlude

Lately I seem to be hearing the sounds of hippopotamus wherever I go. In the bush, looking at films, on tape, on radio – and I have come to the conclusion that their language is quite marvellous; not at all the single grunt-bellow I had been led to believe.

Even on a windy, rain-swept day their resonant, echoing call finds you wherever you happen to be. It may be round your blazing campfire, which could enrage the dominant bull to the fury of a head-on charge. Fortunately for you, he stops the moment you damp your fire down to the point of glow.

Hippopotamus are shy of light, preferring to venture forth during the moonless hours. Considering their short legs and their enormous bulk, their agility on land is amazing. The hippo is a creature of rugged individuality, yet at the same time he is an inseparable part of his close-knit herd, which protects its young against such aggressors as lions and crocodiles.

To gain even a small measure of insight into their lives, you must take time to live within earshot and sight of spoor-print, and even then it takes days to become accustomed to their vocal variations, their explosive moods and to the pattern of their lives. In time, as you listen to their erratic chorus, you will learn to separate the sounds of love and war and the peaceful variations of their honking.

What surprised me, on a sunlit morning as I gazed out over the Shiloweni Dam in the southern section of the Kruger National Park, was the patience and tenderness they displayed towards their young, something I had previously only seen among a herd of elephant. Usually, even the most doting parents will at times lose patience with their offspring. A baboon won't take long before it metes out physical punishment; a lion's disciplinary paw at a feast may injure or even kill an over-eager cub.

On that special day there were three hippos out on land in broad daylight, an unusual occurrence. The largest of these was endowed with a set of fearsome tusks and plenty of bravado, unmistakably a male. The cow, in contrast, was sleek and feminine, keeping close to her calf, a perfect miniature of its elders.

They were communing and I soon realised that the infant's insistent nudges into his mother's flanks had a purpose other than a display of affection. The cow was gradually moving towards water and it seemed as if the calf was trying to stop her – resisting, pushing, yet gradually borne along by her great mass. Was it possible that the calf was reluctant to take its first plunge?

The bull opened his cavernous mouth and bellowed as if in anger, circling his mate once, then pushing at the calf with an impatient movement of his head. The cow blocked his path, laying her mouth upon his as if in supplication and he retreated a little, standing with head turned up, grunting to himself from time to time. The female then advanced upon the calf and began to push it slowly but firmly, step by step, heading towards the water's edge and into the shallows.

At this the infant squealed in alarm and rushed back on to land, only to be persuaded into the water again, this time with more success.

The disgruntled bull, his nose pressed to the cow's tail, rejoined the pair, wading into the dam until all three were immersed and became transformed into one large splash and grunt and ripple.

Over the rise an exquisite male waterbuck stood still as a statue, contemplating the pool. We waited for him to come down, but he would not venture forth until a flamboyant, bristling warthog broke through the scrub and hit the narrow game trail with jaunty aplomb.

Don't Forget the Big Birds

If you want to hear the voice of the ground hornbill you have to be an early riser, even earlier than that dawn herald of the bush, the red-eyed turtle dove, which manages to tune into its world before many other songsters become audible.

There is an atmosphere of unreality about the wilderness at that hour, a receding of nocturnal sounds, a gradual merging with the daytime chorus. Each sound that your ears – or more likely your earphones – pick up, adds to the symphony that is never completed, for each day brings new notes that may be hard to understand or interpret, but that you will never forget.

Such is the boom call of the ground hornbill that reaches for miles, proclaiming its territory which it doesn't mark, but constantly patrols. The only other creature that emits anything like this sound is the giant eagle owl, though the rhythm of its vocalisation is different. The red throat-pouched (the female has a blue-purple patch on her pouch) hornbill's chant sounds like an echo from another world, in total contrast to the appearance of the caller, which is more like a terrestrial turkey than a nimble inhabitant of the air.

From Alan Kemp, ornithologist of the Transvaal Museum of Pretoria who works closely with his biologist wife Meg, I learnt some remarkable facts about this bird, the largest and most sociable of the hornbills, of which there are almost 50 different species. Alan, who has been engaged in several years of field study, is much impressed by the life-style of the ground hornbill. 'Studying this species is more like studying a primate,' he explained.

They live in a stable society, move in groups, observe status, respect and defend their rights within their tightly-bonded existence. They cannot fly higher than 120 metres and do not, as in the case of all other female breeding hornbills, seal themselves into their nest-hole. If the society suffers from female domination, it doesn't seem to have any detrimental effect on this carnivorous, orderly, somewhat bizarre creature. With their large beaks – like pick axes – they prey on such animals as insects, squirrels, snakes and tortoises, moving through their terrain with a somewhat tilting gait on not very efficient feet, nesting in tree- or rock-holes which protect them from enemies, such as leopard. Their beaks have other uses, such as affection display, or making tapping noises on some hard object to attract attention.

Strangely, though indigenous to the bushveld the ground hornbill seems to suffer from bouts of heat discomfort, counteracting this disadvantage by a special shoulder-lift posture, which increases the air-flow under the wings, in this way helping cooling currents to bring down the temperature. Once, long ago, the hornbill was a forest dweller. No longer sheltered by dense, high tree-crowns, he begins to suffer from over-heating at a temperature of 79–83° Fahrenheit.

According to Alan, the ground hornbill can be followed with ease from a Land-Rover. This makes the task of detailed observation very much easier than that, for instance, of the oxpecker who builds his nests in elevated tree-holes and covers large flight areas between his host-animals, such as buffalo or giraffe, which often provide building material.

The Ways of a Serval

Unless the moon is full and the sky is clear of drifting cloud, it may be difficult to be certain that the fleeting vision of spotted skin and furry tail adds up to a sighting of one of those elusive, nocturnal 'lesser' cats about which so little is known. Civet, genet, wild cat, caracal or serval? Was it one of these that has just passed your way along its scent-marked hunting trail?

If your vehicle has a spot-light you might, with luck, get more than just a glimpse of the night prowler, its eyes red or amber or luminous green, depending on the cell structure of the eye, which has evolved to fulfil its special need for optimum survival.

Cats are endowed with excellent eyesight, their pupils opening only a slit in bright light, but widening to accommodate more light during the hours of 'darkness'. Unlike human eyes, theirs are not evolved for detailed vision, a feature most highly developed in the eyes of the hunting owl, which must register exact detail of form to fulfil the demands of its specialised dietary need. The owl's eye lens is shaped like that of a microscopic lens, cut away at the edges, while its rodent prey, perceiving the predatory shadow and not the detail of outline, possesses an almost spherical lens, able to pick up all possible light rays, the distance between lens and retina being very small.

Trying to tune in to that world of nocturnal shapes, sounds and movement is like entering a foreign domain. And yet it seems that our human senses can gradually become accustomed to the shadowed wilderness, so that we may share the lives of creatures diverse and doubtless, as I discovered in the case of the serval cat, of amazing intelligence.

Taller than other small wild cats, the tawny serval stands about 50 cm at the shoulder, its handsome markings easily recognised when, even briefly, it cuts across your vision. Knowing that it is almost as fierce as the pencil-eared caracal cat, I was all the more surprised when my short-tailed, long-legged patient, a wild orphan with four kittens, displayed a pattern of behaviour that proved her immense mental ability, coupled with a particularly powerful maternal instinct.

Set at the entrance of the Nairobi National Park in Kenya, the animal orphanage contained injured, lost, displaced and orphaned wild animals, many of which were re-orientated to resume their previous wild, free existence, if that were possible. Others who stood little chance of any life but that of captivity remained in the little 'zoo' and provided the stimulus visitors needed before going out into the wilds to observe those same animals living out their natural lives.

The serval cat had come in as an orphan foundling, adult and in kitten. I was called in to vaccinate her and her litter against virus enteritis which had recently swept through the feline inhabitants, leaving many casualties in its wake. The caracal and cheetah inmates had presented only the normal problems of restraint, accepted as routine in the course of one's veterinary duties. Finally we entered the serval cat's pen, and one of the skilled animal scouts preceded me to subdue the first kitten by means of a catching-net at the end of a long stick, calling to me to follow and administer my injection.

Moving away from the netted, squirming bundle of fur, he had

momentarily turned his attention to the other kittens, encamped in a narrow space behind their night-house. Suddenly, like a streak of lightning, the serval mother leapt towards her captured young and, picking up the metal rim of the catching-net in her teeth, released the prisoner.

We watched, spellbound, as she pulled the long stick and net towards their sleeping quarters, growling with frustration as the stick got caught against the side of the wire fence, not content until she had dragged it out of sight.

A few who have had other proof of powerful intelligence among animals say this story is nothing out of the ordinary, for why should man believe that only he and his nearest relatives down the evolutionary scale have been endowed with the ability to act swiftly, sanely, and successfully in response to unexpected situations?

Others disbelieve and shut their eyes to the truth of conscious perception, judgement and rational thought among animals.

Is such an act of courage prompted by blind instinct or is it, as I believe it to be, the proof that we haven't yet begun to uncover the remarkable ways of the wild?

The Spotted Cubs

Among nature lovers 1975 will be internationally remembered as the year of the cheetah, a year when the birth of twenty cubs at a private farm near Brits in the Transvaal had turned the tide

of gloom into an era of new hope for the continued survival of the spotted sphinx, faced with near-extinction within its severely restricted environment.

For some years now biologists have become increasingly aware of the urgent need for new and vital field research into the reasons for the cheetah's population decline.

Does the weakness lie with physical disabilities, enhanced by

their forced adaptation to new types of habitat, such as woodland and rock terrain?

Have the growing stresses of their curtailed hunting grounds reduced the cheetah's tolerance to disease? Could this highlight psychosomatic conditions in a creature of such extreme sensitivity and non-aggressive, vulnerable temperament?

Is it possible that the cheetah species is, in fact, on the way out in the evolutionary sense, its departure hastened by adversities imposed by man?

Do these adversities lessen breeding efficiency and fertility, and post-natal infant care?

By the time some of these enigmas are solved, will sufficient cheetah remain so that new findings and new management policies can be related to them?

Who has ever understood the ways of the spotted sphinx, from birth to death, either under remote- or close-contact conditions; and who has attained the knowledge that would enable him to understand the complex personality of these ancient, revered, world-record-breaking sprinters of the plains?

After years of observation and undivided devotion to her task Ann van Dyk, aided by a close-knit team of her own brother and a private Pretoria veterinarian, as well as the expertise, encouragement and backing provided by Dr Brand, Director of the Pretoria National Zoological Gardens, has succeeded in discovering the correct recipe for breeding success. She has worked with the largest group of cheetah ever kept under these conditions, numbering 29 in all – 19 males and 10 females.

'Take one male and one female animal of any breed,' some of my own over-optimistic clients used to insist, 'and you are bound to have unfailing success with reproduction. After all, they are just animals, and among animals their basic instinct takes over every time.'

Nothing can be farther from the truth, as Ann van Dyk discovered very soon after she embarked on her cheetah breeding project. Animals, and especially wild animals, are highly selective in the choice of their mating partners, expressing themselves in many different ways, from gentle to violent, from very subtle to extrovert behaviour.

If partnered with an aggressive, powerful male, Ann explained, the female, though unimpressed, will react only by uttering the submissive call, without resorting to a physical demonstration of antipathy. If an ill-favoured paramour is not the strong male type, she will attack and repel him within moments.

How does Ann select her candidates?

'It is all a matter of tuning in to their kind of personalities,' she explained. According to her co-workers she is, indeed, blessed with an uncanny instinct for knowing exactly who might accept whom, at what time, what moment, and in what manner, something which requires so much more than scientific knowledge.

The finesse of the mating act, the time it takes, the locality in which to introduce the mates, and the post-copulation period, these entail day by day and often hour by hour observation which can only be achieved by a harmonious, non-irritant presence in total empathy with the charges.

In addition to this task, which includes health-checks and careful food control, skilled veterinary examination of male fertility has proved of the utmost importance in establishing which males are most likely to prove successful fathers. This investigation is throwing new light on some breeding failures of the past, since already a number of male cheetah have been shown to be partly, or totally, infertile.

When success came all but one of the females in litter turned out to be hopelessly inadequate mothers, which often happens among felines in the wild.

Inevitably Ann van Dyk added infant care to her long list of duties, hand-rearing five litters born within five months.

Memories of my own wild cheetah patients in East Africa came flooding back as three nine-week-old spotted bundles romped over the note-pad in which I was writing the story of their advent into the cheetah-depleted world.

The World of Insects

When you enter the realm of insects, those 'vague, unconscious, rudimentary, almost nameless little lives', to quote Maurice Maeterlinck, it is hard to know where to begin to look and listen, for our world is filled with myriad, varied forms of whirring, rasping, clicking, droning life. About 700,000 species have been described and every year new types are being discovered and studied. Their total is nearly three times the number of all other animal species in existence on our planet, ranging from single-celled protozoa to ones the size of mice, such as the Atlas moth in India.

The voices of insects probably first filled the air over 300 million years ago, long before the advent of the dinosaur. Yet on modern man, living within a sound-polluted environment, their many-toned calls usually fail to make any significant impact. Worse still, most of us allow the visual world of insects to flit by without registering the wonder of their ways, our untrained eyes more adept at perceiving the giant form of a browsing elephant than the flickering wings of a passing insect.

Fortunately there are insect addicts who devote their time to studying the life-patterns of such incredible creatures as the sociable ants, bees and termites, of sheath-winged beetles, the mystery of the metamorphosis of the caterpillar to the colourful moth. Much of what they see and record is what you can see in your own back garden, as Henri Fabre, the famous entomologist, immortalised the wood lots and sandbanks of France thereby opening a whole new era of insect exploration.

It so happened that Tony Bannister, a Johannesburg naturalist, was taking his leave of me in my Eastern Transvaal lowveld home at the exact moment when a sphex (*Ammophilia tydei*), the 'digger' sand wasp, happened to be flying towards her nest burrow, located dangerously near (for her own safety) the circular driveway next to the homestead. Alerted by the flash of wings, Tony decided to stay his departure for a while, delighted to discover the huntress in the very act of airlifting her caterpillar prey, many times her own size.

Temporarily put out by our presence, she flitted restlessly before alighting at the entrance which led to the nest she had excavated with clawed front feet, her mid and back legs shifting displaced sand behind her.

We watched, spellbound, Tony giving us a running commentary on her actions as she dragged her partly paralysed victim down the angled tunnel to deposit it at the base where she would lay her egg, placing it in such a position that her prey could not inflict harm with its still mobile legs. With the inherited skill of her kind, the sand wasp drives her sting into the thoracic nerve centres (an insect's spinal cord is situated along its chest), allowing for some limited though disconnected movement; the incapacitated caterpillar lying on its side or back is a feast ready for the new-hatched grub, dining on its gigantic host in complete safety.

After depositing the caterpillar the wasp carefully covered the nest-hole with grains of sand, though at another time she might stamp down the surrounding earth by means of a pebble held in her jaws. The astounding tool-using capacity of insects and especially the sand wasp family was first discovered early in the 20th century. Their energy is derived solely from the nectar of flowers, sucked up before the day's work begins.

How do they find their way back to their burrow once they have left for distant hunting grounds? Tony had often observed them at the moment of departure, flying back to their base for just another brief spell of orientation before navigating forth in some enigmatic way, inherited through one microscopic genetic chromosome link.

By now our sand wasp had taken to the air and was busy collecting straws (three times her own body length) one by one. These she dropped neatly over the nest.

Finally, as we stood speechless with admiration at her determined effort to cover up the last shred of evidence of her brood-nest's presence, she flitted off to some distant place, perhaps to bask in the heat a little and to recover from her marathon endeavours, gaining strength for her next excavation, her next hunt, and her next egg-laying.

Lake St Lucia

It was the clarion duet of the fish eagles that jerked me out of an especially severe attack of city blues from which I am apt to suffer after a prolonged stay in the concrete jungle. Many people thrive on city life, which offers both compensations and attractions – depending on the kind of person you are. I, too, enjoy art, theatre and music; indeed at times I long for them and need the mental stimulation they offer, perhaps as contrast to the life I love best which is far removed from crowded sidewalks and the hum of traffic.

On that first early morning at Lake St Lucia, a waterway stretching 72 km along the central coastline of Zululand, and one of the wonders of the aquatic world, I woke to the beautiful refrain of the fish eagles echoing through my bedroom window. Their call to life was irresistible. Within minutes I was up and out, walking on the estuary bank, in the freshness that passes all too soon with the ascending arc of the sun.

I was by no means alone. Scores, perhaps hundreds of fishermen were casting their lines into the cool water that flows out to sea. Patient, but mostly unsuccessful, they obviously revelled in the wildness of the place.

'Until you get mud through your toes', the Director of the Natal Parks Board, John Geddes-Page, had said in a recent address, 'you haven't really been in contact with nature.'

I watched a delicate pied kingfisher vibrating above midstream, his body gyrating, his immobile beak pointing downwards, waiting for the moment to plunge for his prey. He made several attempts to fish while I waited on the bank, fascinated by the perfection of his aerobatic display.

I strolled along the beach, now and then chatting to fishermen, and to women visitors who return to St Lucia time and again. I am told that 160,000 came during that one year.

Are increasing human pressures going to spoil the natural beauty of the environment? Are nature conservation authorities – in this case the Natal Parks Board – aware that many wildlife lovers believe that conservation and milling crowds of visitors cannot ever relate to each other?

This extract from a Natal Parks Board press release provides some of the answers: 'The steady development of public resorts is related to the conservation concept in so far as the reduction of pressure on the more significant natural reserves is essential. South Africans, especially, are attuned to and in constant need of the outdoors with its golden miles of beach, tranquil lakes. Communing closely with nature isn't for everyone, and since there are really those who prefer motor boats to antelopes, why

shouldn't they? The Natal Parks Board nevertheless intends to establish in every public resort a wildlife microcosm . . .'

Gilbert Schutte, Warden of Lake St Lucia, was showing us the beauty of the northern part of the wildlife microcosm of the great lake from the helm of the Park's Hamilton Jet boat, skimming north from Charter's Creek, past vegetation-clad shores and forests, ancient rock-drops and caves that must be marvellous to explore.

The history of this terrain covers a period of 300 million years before man arrived, a time of great geological upheaval.

Reedbuck, the tiny Suni antelope, bushpig, nyala, bushbuck, duiker, red and grey, are seen in the lake-shore forest. Crocodiles, the largest I have ever seen, bask among the bank reeds of the Mkusi River. Hippo abound in great herds, living out their lives in the security of protected sanctuary domain.

Perhaps, above all, St Lucia is best known for its vast population of birds.

'Flamingoes ahead!'

I could see only a faint white line against the deep blue ripple of waves. We passed Bird Island to our right, dense-white with breeding spoonbills, the nest-mounds of grass and sticks already empty of chicks, which moved below the long-legged adults, their grunts and squawks reaching us as we passed by.

As the white circle ahead grew larger, Gilbert Schutte slackened speed and cut engines. Very suddenly the wind dropped; within moments the water top was smooth as glass. I set my sound-recording equipment in motion and pointed the long-range microphone in readiness while Gilbert plunged the boat forward once more, then stopped again.

I could hardly believe my eyes. Before us thousands of greater flamingoes communed in honking chorus, a great, gleaming mass of wings and necks in snow-white, tinged with softest pink, their black-tipped bills held high. Like a ballet in slow motion, the flock separated and streamed into a leisurely take-off, skating across the water for the final lift, their black and red regalia flashing with each rhythmic wingbeat. They circled and landed again, a little distance away, as if they had moved only because they preferred a small measure of privacy.

Listen to a Crocodile

Have you ever really looked at a crocodile? Or listened to its voice?

My companion, to whom I addressed the question, regarded me rather pityingly, but I insisted that he do more than just brush me off with a shrug of the shoulder, as if to say: 'I always knew that the African sun had scorched your common sense away long ago. And now, alas, it is too late.'

What is there to look at, he asked, but a torso of green-black scales, glinting evilly on the river bank, waiting to kill. And as for its voice, who ever heard of a croc expressing itself in sound, except, of course, in that first moment of birth, from egg to life, when all creatures, even those ugliest of saurians, appear entrancing.

Not so long ago I too was prejudiced, uninformed, as guilty as the next man of maligning this magnificent creature which, only now, is coming into his own through research and patience, and that only in very few parts of the world where crocodiles still roam.

After I had met Tony Pooley, Warden of Natal Parks Board Crocodile Breeding Research Station, now situated at St Lucia in Zululand, my view of the croc tribe altered profoundly, for under his guidance I learned to look and listen to his charges with new eyes and ears. Several visits in the past had proved memorable, but none more than a recent examination of his new croc pools, established when the internationally famous breeding centre had to be moved lock, stock and barrel from Ndumu to the Estuary Reserve.

It was still cold, though near nine in the morning, when Tony threw the first piece of meat and bone to attract the crocodiles to the fence-line for close observation and sound recording. We did not have to wait long. Responding to the special pitch of his voice and the lure of food, the great scaly patriarch moved towards us and close to the fence, his torpid body astonishing Johannesburg cameraman Tony Bannister and myself with the incredible turn of speed and agility as he flashed his jaws to gulp his portion. Crunching it very audibly he slowly ground down the meat, jaw upon jaw, top upon bottom, with no lateral movement such as a lion or dog would use to chew from side to side. The croc crushes to size, then swallows and as he does, a growling, truly dinosaurian-type belch escapes from the cavity of his digestive tract, gas being displaced by bulk as he stretches upwards, forelimbs rigid, head high, until the growl becomes a long-drawn hiss that ends in a sigh.

I had already recorded the anger-growl of a crocodile, a true laryngeal sound, and heard the awesome roar of contesting males captured by Lourens Fourie, of the South African Broadcasting Corporation, who taught and encouraged me so much in my wildlife radio programmes. Best known of all is the yelp-squawk of the newly emerging infant, but there are doubtlessly many other sounds of communication which have not yet been recorded or interpreted.

His meal complete, the male returned to the prone position, immobile but for the flickering of his green-slit reptilian eyes and the barely visible and, to us, quite inaudible rhythmic heaving of his flanks with each respiration. I began to study the pattern of his scales and scutes and wondered at the symmetry of the close-knit armour plating.

Another, and then another smaller crocodile advanced, drawn by the food which Tony Pooley, once in a while, pitched over the fence.

I wondered how the new adolescent arrivals dared come so close to the elder. Then one of them, venturing too near, was pursued out of reach as the large male catapulted into sudden action with a flash of lashing tail that became a rush of water as he chased the young croc out of sight. It was a show of fearful speed and power and constant alertness which unless seen might never be suspected.

On the opposite bank the matriarch basked in the sun. It was she who had demonstrated the amazing maternal behaviour of her kind, carrying her new hatchlings, which she had jealously guarded during incubation, down to water in the elastic gular pouch of her throat.

In the wild state even her presence would not have prevented the advent of a multitude of croc-infant predators, drawn by the high-pitched explosive yelp. Eagles, python, mongooses and, above all, the monitor lizard, take a heavy toll of the brood from the moment of birth. Once released from the mother's pouch into water, the battle for life has by no means been won. Even then the mother croc, in this instance assisted by the male, would for several months continue to fight off any intruders that might threaten her young.

Who would have believed that the ancient, cold-blooded crocodile, once believed to be cannibalistic, would show behaviour traits as highly evolved as those of many mammals? Perhaps, if you looked at the world around you, you would find that their code of motherhood ranks very high indeed.

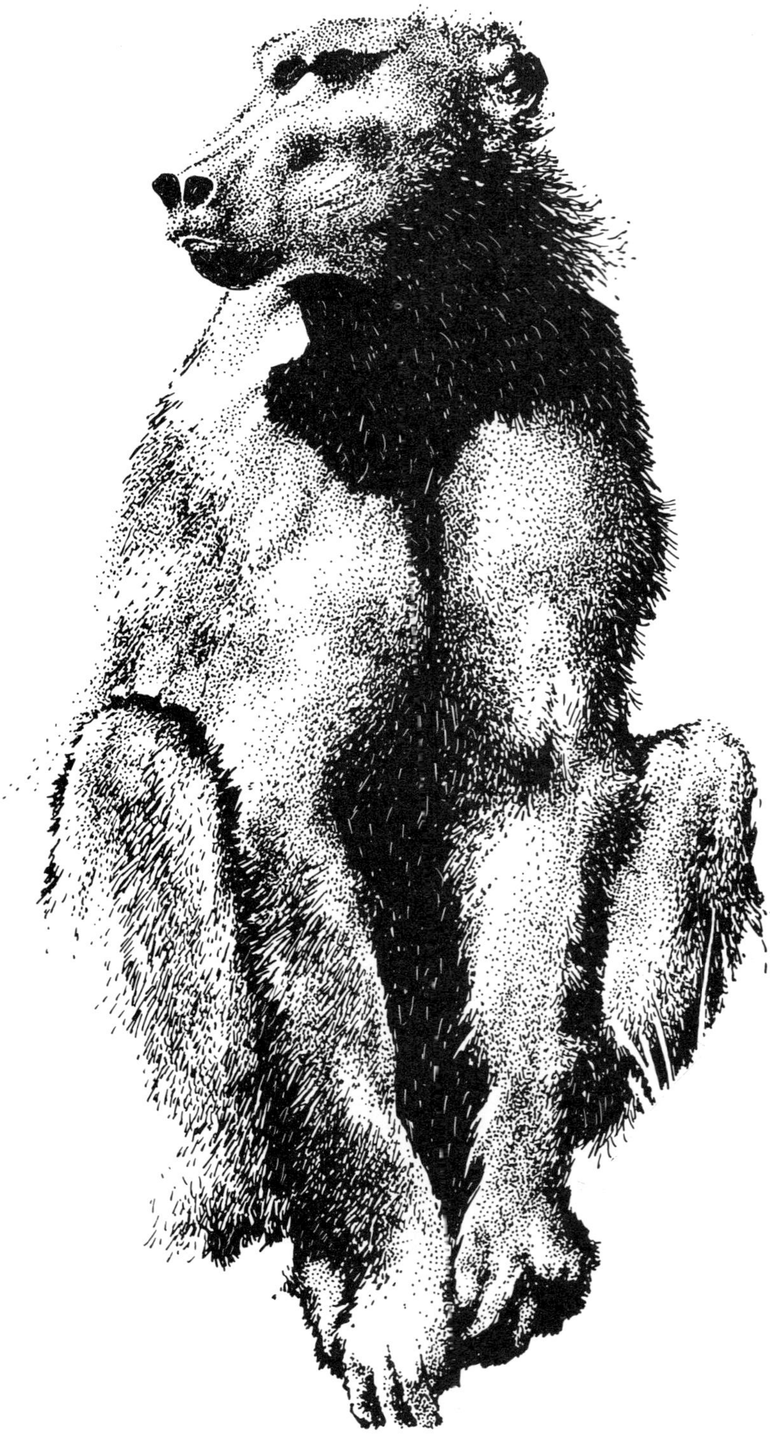

Instinct or Intellect

It would be difficult not to ascribe certain character traits to the wild animals one knows best, especially when one has spent years observing them under natural conditions. Those years, too, make one realise that innumerable unpredictable reactions are set off by unknown factors and account for astonishing behaviour patterns that you may witness once, and perhaps never again.

Consider the recent confrontation between a leopard and a porcupine, reported in the National Parks Board publication 'Custos', August 1975. How was it possible that a fully-grown, presumably experienced hunter, such as this leopard, was not aware of the spiny nature of the quilled porcupine?

Was he a leopard of stunted intellect and instinct, an exception amongst the most cunning of all spotted wild cats, who are renowned for their judgement and wisdom? A leopard, handicapped by quills which he might not be able to extract, could eventually die of starvation.

What, then, is intelligence? Can one assume that it is the sum of accumulated experience, or does the individual inherit intelligence?

Primates usually show highly evolved behaviour traits which, to my knowledge, seem more predictable than some of the other mammals. I have always been fascinated by the recognition of death in animals, and at the other end of the scale, by their show of mischievous humour. Creatures that give clear evidence of either of those surely possess something more than just instinct.

A chacma baboon gave birth in captivity. When the baby was a few weeks old it was injured in an accident and was removed from its mother to relieve its suffering and, if possible, to save its life. For three days the mother hardly ate, calling day and night.

On the third day the baby died, in spite of treatment. The body was placed on the ground before the mother. Her restless excitement at once subsided. She approached the body, making chacma sounds of endearment and touched it twice with her hands. She then put her face close to the back of the dead infant, touching its skin with her mouth, at the same time moving her lips. Immediately afterwards she got up, uttered a succession of cries, walked to a corner and sat down quietly in the sun, and took no more interest in the body.

As for animal humour I have always enjoyed the tale of the two bull elephants, told by that great conservationist Vesey Fitzgerald, once Warden of the Ngurdoto Game Reserve on Mount Meru in Tanzania. I quote the version in *The Flying Syringe* (Geoffrey Bles, 1970) by A. M. Harthoorn:

'An electric fence in Tanzania was inspected by bull elephants who soon realised that delicate trunks suffered disagreeably when in contact with live wire strands. Two large bulls stood, head to fence, apparently in deep thought, but with a space between them. The space was soon filled by a rubber-necking young bull intent on finding out just what was going on. As if by pre-arrangement, the large bulls seized the youngster by "the seat of his pants" and pushed him, in spite of protesting screams and stiff-legged resistance, right through the fence; the rest of the herd was then quietly led through.'

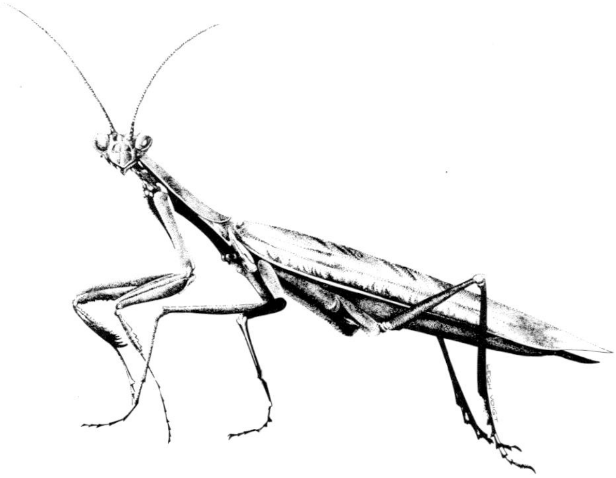

Sounds of the Bush

Within the confines of the wilderness the drama of life and death is played out at every moment of the day and night. If you are fortunate you may witness some incident or confrontation, but more often wild animals move under cover and out of sight, their presence only detectable by a trail of footprints, or a heap of dung, or chew-marks on a tree-bark, or hair that has become embedded on a favourite rubbing post.

Above all, they reveal their whereabouts through sound of many kinds. Sometimes the wing-claps of a lark turn your attention skywards, or the tap-tapping of a tok-tokkie beetle brings you down to earth. Bird song, infant contact calls, playtime, conflict, these are being enacted perhaps a few metres from where you happen to be.

I remember the first time I heard what sounded like a rifle shot and traced it to two male adult giraffe, fighting in earnest, venting their emotion, perhaps their jealousy of territory or a mate, by engaging in the spectacular 'necking' routine.

You may have seen the long-necks making love, twining and swinging back and forth in almost slow motion. Even then it looks dangerous, but I am told that this kind of neck manoeuvre is equivalent to a caress, the prelude to eventual mating. When giraffe bulls fight in earnest the mood of the contestants is very different, their facial expressions indicating intense stress, their tails lifting and curling sideways with mounting tension.

On this particular occasion the tempo of movement altered noticeably; the impact of the head and 'horns' (not true horns, but exostoses from the skull) causing tremendous thumping of the adversary.

I have often seen deeply-scarred individuals, no doubt the result of repeated neck-sparring, but I have never heard of giraffe actually killing each other in combat.

At times it is the sky-climbing vultures, drifting over the scene of carnage, that tell you what you cannot see. You stop, switch off the engine, listen to the crashing, growling, screaming sounds emanating from a hidden thicket in the bush, and guess that a lion pride has claimed a victim. A roar and then a yowl might indicate that a lion, leader of his pride, is disciplining a cub.

The insect world, especially, remains hidden to most of us, unless we take a close look. Small sounds with significant meanings may give us a clue, or we may see a vividly-coloured leaf or stick suddenly come to life, like the stick-insect that rattles its staccato warning at the most unexpected moment; or the sudden rustle of the wings of the praying mantis, so peaceful in appearance that in Provence it is known as *lou Prègo Diéou*, which means the Prayer-to-God. J. H. Fabre describes this deadly creature in *The Insect World*:

'Her long, pale-green wings like spreading veils, her head raised heavenwards, her folded arms, crossed upon her breast, are in fact a sort of travesty of a nun in ecstasy.

And yet, she is a ferocious creature. The work-yards of the various digger wasps receive her visits frequently. Posted near the burrows, on some bramble or other, she waits for a chance to bring within her reach some of the arrivals, forming a double capture for her, as she seizes both the huntress and the prey.

Her patience is long put to the test; the wasp suspects something and is on guard; still, from time to time, a rash one gets caught.

With a rustle of wings, half-unfurled as by the violent release of a clutch, the mantis terrifies the newcomer, who hesitates for a moment in her fright. Then, with the sharpness of a spring, the toothed forearm folds back on the toothed upper arm; and the insect is caught between the blades of the double saw.

It is as though the jaws of a wolftrap were closing on the animal that had nibbled at its bait. Without loosening the cruel machine, the mantis gnaws her victim in small mouthfuls.'

Night of the Full Moon

Most of us who had come to Londolozi Game Reserve that day were utterly exhausted by the time we reached the platform mounted in the great, sheltering sycamore fig tree; it was then three hours after sundown.

The light of the full moon reached us through the canopy of the tree-crown, adding a touch of magic to our retreat. It is hard to resist sleep after a long day in an open vehicle. As I watched my companions snuggling into their sleeping bags, I fervently hoped that I would remain awake for a little longer, for I wanted, above all, to tape-record the calls of this night of the full moon and, if possible, the changing of the hours from darkness into dawn.

I made notes, remembering the day, retracing my steps hour by hour, writing to the purring song of a Scops owl, a truly nocturnal hunter feeding on insects which its well-adapted eyes perceive in the shadows of the woodland.

How totally different it was to the great diurnal raptor known as the bateleur eagle, so named because of its slow-rolling aerobatics. We had sighted several of these (very aptly chosen as the emblem of Londolozi) speeding across the horizon, wings angled backwards, searching the ground below for food – perhaps a reptile, or carrion, or a lesser bird.

In flight the underside of the wings shows pure white with a black trailing edge. There is surely nothing more magnificent in the world of birds than the bateleur's coal-black and chestnut outline and the brilliant red, bare skin of its face and legs.

About the middle of the day we had come upon a recently killed female adult kudu, guarded by three lionesses blooded from chin to paw, mothers of eight cubs, varying in age from eight months to six weeks. There was a scorching quality about that hour and we did not have long to wait before one lioness left her patch of shade to move the antelope, exquisitely graceful even in death, out of the sun.

The lioness looked at us hard and long, then moved forward

to straddle the body, grasping it behind the head as she slowly but surely performed the amazing feat of lifting and dragging the cow 50 metres or more out of the circle of light, and then farther into the shelter of the thick thornveld bush.

I could not but admire the rippling power of the spinal muscles in tension and the certain, measured technique of the manoeuvre. How much of the kill would remain for her to consume? The lionesses and cubs were closing in at that very moment. Perhaps, as so often happens, the pride leader might come in too and lay claim to his share. Each lioness could eat up to 20 kg of meat, each cub would take according to its size. There might only just be enough to fill all their bellies for a few days.

At this moment hyena shadows moved across the clearing below our tree, following where, a few moments before, a genet had passed – flowing, rather than trotting, its tail markings catching flecks of light.

I switched on the recorder to capture the hyena contact call. The whoop rising to a high crescendo, then descending into a minor-scale lament. After a time I began to distinguish their individual tones and expressions, recalling how Simon Bearder, the biologist of Pretoria University, had explained hyena vocal recognition and response in relation to behaviour patterns.

Insects, a nightjar, a pair of eagle owls, a lion rumbling like a distant storm, the sounds I loved were beginning to lull me to sleep, so I ceased to resist, until three hours later, the morning doves gradually brought me back to consciousness and to the freshness of another lowveld day.

Umfolozi Trail

In the flickering firelight, which we could see through the open tent flaps, the giant sycamore fig seemed to radiate a kindly, sovereign presence, sheltering and dominating our encampment on the banks of the 'white' Umfolozi River. The great tree must be very old, I thought; perhaps 400 years or more? Its sturdy branches, one of which measured over seven metres, reached out horizontally like the protective arm of Father Time himself; its gnarled roots seemed as indestructible as the foundations of an ancient cathedral.

One lion, then another, rumbled and roared from across the river. Perhaps I imagined it, but it seemed that my stretcher shook a little with the impact of their nocturnal chorus. A short burst of whining jackals sent a shudder of sudden fear for the hunted beasts down my spine. Here at Umfolozi there is plenty of fodder-on-the-hoof for the predators.

The exquisite nyala, startling in their extreme sexual dimorphism, abound among the scrub thickets along the river, displaying their finely decorated, slender bodies that freeze into total immobility as they stop to gaze at you with deep-brown, soulful eyes. Resembling the bushbuck in many ways, the nyala bull carries a shaggy fringe of hair on its underside, the only antelope in Africa possessing this feature. Blue-grey with a tint of brown, this handsome, browsing creature boasts, like the kudu, a chevron of finely-drawn white between the eyes, highlighting rather than hiding its head-on approach.

A hornless female, bright chestnut with white side stripes, without chevron or undercoat, had seemed shy and elusive that morning, as if unused to human admiration or disturbance.

The white rhinoceros cow and calf we had met almost face to face in the afternoon had not taken kindly to our intrusion into their woodland retreat. Two women among nine men, Panorama reporter Wilna Russeaux and I had taken to the only climbable tree in the vicinity, leaving the chivalrous men to take cover as best they could while the trail-leader, Warden John Forrest, guarded our safety every inch of the way.

After repeated mock-charges, the rhino pair finally allowed us to continue on our way in peace, but they had alerted us to the fact that here in the unspoilt wilderness one meets Nature face to face and accepts the element of danger as part of the experience.

We stopped to wonder at a polished stump of tamboti, a rubbing post for rhinoceros that, perhaps, also serves as a rendezvous. Dung heaps old and new told of great numbers of rhino and of their food habits. The black rhino's dung contains sticks and other processed browse; the grazing white rhino passes ground-level byproducts of grass, reeds and shoots.

The fresh scent of animals and blossoms and, as we approached the river, of water and damp sand, soon obliterated the memory of city pollution which, for a time, can dull one's perception. The breeze carried the unmistakable aroma of buffalo. A herd had passed our way some hours before and could be seen drinking along the river shallows as we descended the rocky slope.

That evening, the first of our trail, as Wilna and I sought the privacy of a river-clump for our ablutions a startled reedbuck noisily broke cover at our feet. Having put some distance between us, the buck checked and took stock, while overhead, a yellow-billed kite flew towards the cliff-face that the Zulus have named Mohboehni, where the wind is like the breath of the puff-adder.

Looking back to those days in Umfolozi it was the early morning when I walked to the river, examining the evidence of life on the sands, that had been my favourite hour.

Lion had prowled there during the night, their spoor visible next to the sharp fine imprint of the water-side birds; there were kudu, impala, a small cat-print – perhaps of a genet – buffalo, wildebeest. A great multitude had moved over the sand, claiming their water-rights while we slept just above them under a grove of fig trees that is also part of their world.

Island of Orphans

Had I not known Toni Nutti, now an almost legendary figure whose five-star inn on a remote island on the Kagera River in Uganda has delighted long-distance travellers for scores of years, and had I not seen her island menagerie, I might not have wholly believed the many tales of her wild animal orphans – especially not the epic of the stranded baby hippo and the klipspringer.

The daughter of an Italian prospecting engineer, Toni, once a renowned beauty, not only knew how to entertain her guests but possessed great empathy with the many wild creatures brought to her. Once she had restored them to health she often returned them to the wild. Even young elephants, still dependent on liquid nourishment, flourished under her care, for she seemed particularly suited to satisfy the psychological yearning of small tuskers, which often die when removed from their mother.

If only I had returned to her inn the following season – as I was invited to do – to record the story of her life in Africa and especially the secret of her success in rearing wild animals. Alas, before I left East Africa Toni Nutti had died, taking with her years of experience built on patience, an uncanny instinct for nursing the wild and her own, well-remembered dictum, which she explained in simplest terms: 'Live with it, sleep with it, talk to it, make physical contact, don't let it out of your sight – very awkward at times – but it must be so, for with a baby elephant you must be its mother whose trunk reaches out every few seconds. And then,' she had added with a glow of remembrance, 'even a little almost-dead elephant calf, starved of mother-love and mother's milk, will live. I know, I have done it.'

What indeed had Toni Nutti not done in those years when her fame had spread to the four corners of East Africa. Unfortunately for elephant orphans the bridge to the island had been washed away in an especially heavy flood and was never rebuilt.

Instead, a hazardous chain-pulley boatswain's chair transported visitors across the swirling waters, providing a very novel, thrilling experience as the cage was drawn across at an unhurried pace. A small elephant could well be ferried to the island. If it thrived, an irreversible problem arose since the pulley could load no more than two people at a time.

I have often asked, but no one could tell me, how many animals Toni Nutti restored to life. However, everyone who visited her seemed to know of Pepi, the klipspringer, who had come to her as an infant foundling.

On the evening of our visit Pepi had sat upright in an armchair listening to the tale about himself, or so it had seemed. I had stretched across and stroked his bristly coat, not realising that it would come away at my touch; those pithy, stiff hairs had once been used as stuffing for saddle panels.

I looked at the amazingly small feet, almost rectangular in shape with long, narrow soles, and was reminded of the many times I had seen klipspringer leaping effortlessly from boulder to boulder on rock-gripping vertical hooves, moving along as if their legs were set on springs, their alert, compact forms blending into the rock face.

Did Pepi miss his homeland? Would he eventually find a mate, perhaps another orphan?

I clearly recall the evening when, to the constant murmur of the mighty river, Toni Nutti related this almost unbelievable story of animal kinship.

Some months before, late at night when her guests had retired to bed, Toni had noticed that Pepi was missing from his customary place. After a thorough search of all his favourite haunts, Toni had drifted into uneasy sleep, only to be woken some hours after midnight by a strange, persistent tap-tapping on her door. On opening up she had been amazed to find her klipspringer in the doorway, standing close to a large shape which at first she could not identify.

On closer examination the candlelight revealed a sad-looking, very weak hippo, its face shrivelled and worried like that of a displaced human infant.

The unlikely pair had just stood there, looking up at her, Pepi expressing his impatience through several short, flute-like whistles, and hoof-tapping on the floor. Taking the hint as soon as she could gather her wits, Toni prepared a milk formula for the orphan. Within two days of regular bottle-feeding the little hippo had found new strength and was beginning to disrupt the Nutti household in no uncertain manner.

Having discovered what appeared to be its calving site in the back garden cabbage-patch, Toni decided to take a chance and attempt to return the baby to its mother, for she had noticed a lone, large hippo near the island on several occasions.

What a sight that reunion must have been as the herd accepted the human-scented calf without question, an event that will be remembered for many years in that far-off region of Uganda.

Fish Eagles' Haven

Interlocking, swaying, dense canopies of gold-stemmed fever trees and high above them, the ringing chant of amorous fish eagles, sounding their mating call every few minutes.

I lay on my back cushioned on the soft, thick mat of wild grass, drifting back over the years when Kenya had been my home, and Crescent Island on Lake Naivasha had been one of my favourite haunts. It was hard to believe I was back in East Africa, feeling no strangeness but only a sense of belonging and joyfulness at having been able to return, even for a short few weeks.

I timed the eagles' vocalisations. Every six minutes, or less, for forty minutes they chanted into the morning, flying across my field of vision so that I hardly needed to move my head to follow their bold flight pattern against the sky.

A watchful male waterbuck moved into the glade and stood still and erect, his backward curving horns deeply ringed with the passing years. He flared his nostrils as he caught my scent, eyeing me for a long time, perhaps wondering at my immobility. Here, on Crescent Island, man is the only predator and at that an illegal one since this land was declared a sanctuary some years ago. When my friend and hostess, Jean Hayes, purchased the island in 1964, only a handful of Thomson's gazelles had survived many years' slaughter.

Now, after eleven years of protection, herds of antelope grace the waterside fields, lords of their universe which, nevertheless, must be constantly patrolled against poachers. They steal in from the mainland in boats and at low water wade across and enter the swamps and thickets.

Within the few years I have been away drastic alterations in the lake habitat have wrought profound changes in the life-style of many of the lake's inhabitants.

Since fish-poaching has affected the fish eagles' and other birds' food source, these raptors have successfully remodelled their feeding habits, preying largely on the prolific 'Louisiana' crayfish, imported years ago by a well-intentioned farmer and which now run riot in several lakes.

On the second day of my sojourn on Crescent I caught sight of an exquisite, thick-tailed marsh mongoose, streaking in and out of sight just before dusk. Its streamlined, almost black-furred

body seemed jet-propelled as it plunged back into marsh and water, moving along its own trail of tell-tale pink-tinged dung heaps that reveal its main diet of crayfish, though not long before the mongoose had fed mainly on birds and fish.

Will this new dietary routine eventually alter its way of life?

Hunting crayfish requires less agility and stealth, less camouflage and cunning. Will its body structure undergo a change, or will the mongoose remain nimble and fleetfooted, equally at home on terra firma and in water, adaptable and hence less threatened at a time when human encroachment, even on sanctuary land, poses new and constant hazards both to plant and animal life?

I got up and wandered to the south-western part of the island where a herd of waterbuck grazed on the lush vegetation within sight of Longenot, that brooding, not-so-long-extinct volcanic mountain. The last eruption had occurred within 10,000 years, though, according to an old Masai tribesman, there had been 'smoke on the mountain' not much more than fifty years ago.

Two waterbuck separated from the herd, a male and a hornless female, and stopped feeding. The female, very much 'on heat', walked stiffly, enticingly, a few steps in front of the bull, her tail lifted, her stance alert, taking a sideways glance to see if her consort was following. The pursuing male caught up with his mate from time to time, sniffing and licking, then drawing back his lips as he raised his head in that characteristic display of male appreciation (known as 'flemen') which I had previously seen in sable and lion during the pre-mating display.

I wandered towards the herd at a slow pace, entranced by the high-jinxing exuberance of the Thomson's gazelles playing and pronking without a care in the world.

Standing on a vegetation-covered rise, two doe-eyed, long-snouted Kirk dik-dik stood silhouetted against the blue lake-top. Not more than 40 cm in height, perhaps 5 kg in weight, they looked as dainty as elves conjured up from some fairy tale.

As I lifted my binoculars to examine them more closely they sped over the top in a flash of grizzle and fawn and slender limbs that barely touched the ground.

Man among Lions

It's been four years since I last visited George Adamson, whom I believe to be one of the most remarkable personalities of our time. More sage-like than ever in appearance, his silver-grey hair long enough to protect the nape of his neck from the scorching sun, he hasn't altered very much in appearance and certainly not at all in manner or in character.

His shy, quizzical smile and self-effacing manner, his ability to listen as well as teach, his special quality of tranquillity which cannot but rub off on human and wild animal friends, these facets of his character are much in evidence, in spite of the many tribulations he has passed through since we last met.

No longer in Meru National Park, north-east of Mount Kenya, where the first pride of lions had been successfully returned to the wild, he lives in a remote region some hours' distance from Garissa town, in a hostile thorn-tangle habitat of haunting, untamed beauty.

Kora, named after the rock-shape facing the camp, has, after years of anti-poaching control, of road building and stringent conservation become a nature reserve in its own right: 500 square miles of wilderness, its Tana River frontage forming a lush, game-rich haven.

Tony Fitzjohn, George's able assistant, drove us from Nairobi to Kora. Patti Ferguson, an American friend who travelled from New York to join me on this visit, had come to renew acquaintance with George, a man she much admired for his determination to fulfil his chosen task.

In the time that remains to wild Africa, George believes preparations must be made for the days when lions may be threatened with extinction and man will need to know how best to return them to the wild. 'Not only the lion, but such predators as tigers in India. Perhaps what applies to the African lion and his ways will apply also to other wild cats.'

I had returned with some misgivings and apprehension, for rarely does one find people and places unchanged. Yet here, world disaster and conflict had made no impact. It was almost full moon and the sky was brilliant with innumerable stars. I felt totally removed from the world I had left behind: Nairobi, Naivasha, South Africa and my own beloved lowveld.

Erigumsa, George's tall, lean Korakora servant, brought soup and home-baked bread which we ate to the accompanying chorus of lions. Our host explained that at times his group of eight separate into two units of three and five and may not reappear at camp for days at a time. As all are young and not yet totally bush-seasoned, George and Tony keep constant watch. Two lions have been fitted with 'collars' which transmit their movement-range to a base-camp receiver. To keep track of the ways of lions radio-monitors are invaluable in dense country such as this.

Leakey (named after Louis Leakey), nearly three years old, is an abandoned cub adopted by George. He and Arusha, who came from the Rotterdam Zoo, both wear collars and seem to make no objection.

Then there is Freddie, who was picked up by a game-warden, and Daniel, the eldest, and Oscar and Kora (cubs of the lioness Lisa who disappeared), almost three years old.

Gigi and Growlie, about 17 months old, were sent to George from the orphanage of the Nairobi National Park.

Both the lionesses and Freddie, the affectionate, handsome teenager just sprouting a mane, were there to welcome Tony Fitzjohn. We watched their display of comradeship from the inside of the high fence, heeding George's warning to take every precaution for safety, especially as some of this group are unpredictable.

We planned to stay at Kora for seven days. What an immense amount there was to learn.

Kora Oasis

In George Adamson's camp the pace of activity is dictated largely by season and temperature. Even in October the hours between noon and four in the afternoon are hot and oppressive, which means that the cooler part of the day must be used to the full.

Nights at Kora have something magical about them. Although the palm-roofed living huts are high and airy I preferred to take my stretcher out into the open, especially as I hoped to make a sound-record of life in and near camp: the constant buzz, the roar and rasp of bush-time melodies.

For me, the post-midnight hours of the first day were the most exciting of the whole week, perhaps because I was in no way prepared for what was to come. There had already been a number of nocturnal but sporadic roaring cadenzas, which, George explained, were simply conversations between lions such as Daniel and Arusha, intoning their 'all's well with the world' refrain across their domain. At four in the morning the five-strong pride really came into vocal action, with Freddie and Arusha very much in the lead.

Their pre-dawn unison lasted for up to six minutes, and gave me the feeling one might get when listening in on a conversation between friends who share a love of life and have no bones to pick with anyone. Speaking of bones, there had been quite a few left around near the camp fence from a previous feast and if there had been any fleshy remains it had not taken the local vultures long to complete the cleaning process.

White-backed, hooded, Egyptian and palm-nut vultures had been at it with their usual aggressive gusto when suddenly a huge, dark, lappet-faced vulture, (sometimes known as the 'King' vulture) entered the arena, the lesser vultures ceasing their eternal squabbling and hissing at once, while the new arrival took what was left of the pickings.

It was the two youngest lionesses, however, which caused quite unforeseen problems, expressing their emotions in the most remarkable sound-pattern, which I was able to capture on tape. Growlie, 17 months old, could by normal lion standards be considered a sexually immature teenager, yet during the past two and a half months she had been showing all the symptoms of a demanding sex-kitten, selecting young Leakey as courtier. Was Growlie suffering from cystic ovaries, we wondered? Was she exhibiting signs of nymphomania which might only be alleviated through surgical procedure?

We discussed her case at great length that morning over breakfast, wondering why, even in the wild, life is beset by quasi-human syndromes that come and go as mysteriously among animals as among man. If Growlie continued this way for another six months, expert medical-veterinary assistance would have to be sought.

Gigi, 14 months of age, also suffered a clinical problem. Her left hind leg, which had been broken through the tibia, had been expertly plated by a Nairobi veterinarian, but had not yet undergone sufficient healing for her to be released back into freedom and rock-acrobatics.

After a brief post-operative return to freedom, I suggested that she be confined within a campside enclosure where enforced rest, short sprints and fence-pacing soon strengthened the injured muscles of the leg, though she frequently protested against her confinement.

George had already mentioned Kora's bird inhabitants and I vividly recalled the bird haven at the camp in Mugwongo, the lion release base at Meru National Park. Here, at Kora, it was no different though I was quite unprepared for the extent of the bird invasion which began just after dawn. First came the vulturine guinea-fowls moving up to and over the fence from their nearby thorn-bush roost.

Their explosive morning incantations are hard to describe in words since they seem to vary in tone and frequency according to age and occupation, sometimes monotonous and brassy, at other times cackling, then again, more harmonious. They explored their day-time haunt for a short while, then began to feed on the grain put out by George in special narrow troughs under the leaf shelter of the main hut. The vulturine guinea-fowl is finely decorated with violet feathers interspersed between grey and white speckles, the cock bird distinguishable by a rudimentary spur.

Next came the fan-tail ravens, five jet-black bush-comics which leave their cliff-roost an hour or so after the guinea-fowl. Their croaks and cries and fascination for bright and sparkling objects can be something of an irritation as is their inability to know just when they're not wanted. Making day recordings in camp with George proved to be a precarious affair, for more often than not a black raven would perch behind my back and totally disrupt a recording sequence with raucous screams and sharp-edged nudges. George reprimanded them but only in the gentlest manner, welcoming all living creatures equally in and near his camp.

And so it was with the waxbills and weavers and starlings and yellow-throated sparrows and the lizards that make their home at Kora, dozens of them, consuming hordes of insects and every now and then one of their number falling prey to a ravenous raven patrolling the little oasis of peace that George Adamson has created in the middle of nowhere.

A Moment in Time

Early morning forays into Kora Reserve frequently end up along the shores of the great Tana River, that magnificent perennial river of life, curving its way between the lush tropical riverine vegetation.

George Adamson loves the river and its inhabitants and he particularly enjoys sitting quietly, watching and listening to the world around him. He is as familiar with plants and fish and birds as with the large animals that seek to quench their thirst in the cool, flowing stream or that feed upon the fruit of the palm trees lining the sandy banks.

Oh, those magnificent doum palms, silhouetted against the morning sky of azure blue and drifting cloud, their finger-like fronds reaching so high that you must crane your neck to see the rich-brown nut clusters, favourite food of high-climbers such as baboons and monkeys, who by shaking them down provide food for the rhino to chew upon.

Elephants also relish the ivory-like fruit which is said to have a high nutrient content. In times of austerity George drank the cocoa-like brew derived from the ground-up husks of the doum palm nut, in place of tea and coffee.

Those same nuts saved the life – so the Warden of Meru National Park told me years back (Meru Park adjoins Kora to the west and north, separated by the river) – of a 'shifta' Somali terrorist wounded by Kenya Parks riflemen, and left behind when the fleeing rebels got away. Two weeks later, hovering vultures and hyenas led the way to the wounded man whose shot-up leg was beyond repair, but who had survived solely on the fruit of the doum scattered around him which he had cracked open on sharp rocks, dragging himself to the river for water.

On our first morning at Kora George took us along the river road which had been literally hewn out of the near-impenetrable thorn cover by George's brother Terence, an expert road builder who will not take no for an answer and who possesses the ability to inspire his men to keep going under adverse conditions which many others could not endure.

We found him at the end of the track more than two hours' drive along the Tana, where he had struck a tented camp which, as need dictated, had to be moved from time to time.

This area had, until recently, been a poacher's paradise. Yet now it is unusual to see a 'poacher's platform' of sticks well above man height, placed within the strong branches of a tree from which the thief commits his crime, usually with twined loops of reed and grass.

Leopard have been poached out altogether. Given time and funds George hopes, one day, to return leopard to the area, stocking it with 'stock-thieves' that are trapped on farmland or near human habitation.

I sometimes wonder whether Kora will (in spite of its easy access for Somali raiders) survive these times of strife and confrontation by virtue of its singular mission. There is something very special about Bwana Simba, the lion man, endowed with a certain aura of uniqueness and mystery that will afford him protection in days to come.

On that very first morning with George our main task had been to check on his three most independent charges who had already mastered the art of hunting and seemed to be killing for themselves. George stopped the Land-Rover at frequent intervals to examine spoor. I shall never really understand how it is that he can read them so accurately, know exactly when the spoor belongs to one of his own clan, who is going where, and sometimes take a guess at why.

Not long after he had spotted their footprints we found all three lions on the remains of a baboon, not a very generous meal for a small pride of large, growing cats. But George had guessed that their stomachs might not be full and had brought meat supplement to make up the shortfall, which they consumed with great gusto, in a sheltering glade of trees, for it was mid-morning and already very hot.

Among the lions rehabilitated in the past, quite a number had died or gone missing. One cub, Katrina, had almost certainly been taken by a crocodile when following her elders, Boy and

Christian, at a river crossing. Christian has not been seen for over a year, and several others have been killed by wild lions. Even so, those who had survived made a life of freedom in their new domain, amply demonstrating that it can be done. George has proved that an ex-captive lion instinctively knows the ways of the bush, provided he is given guidance and encouragement at the start, and is not cut off too early from a source of food on which, at the outset, he must necessarily depend.

As mid-day approached we sought the cool of the tree-shaded river bank, listening to the sound of the rapids bubbling over rocks. George sat on a log of a fallen tamboti tree, thinking aloud, allowing us to share his thoughts. A hippo grunted from round the bend, a goliath heron squawked hoarsely as, with slow deliberate wingbeats, it flew up and across our line of vision. A wood ibis stood fixed and attentive, no doubt waiting for his next fish-catch.

Across the bank, among acacias and poplars clad in delicate light-green foliage, bee-eaters flew up, dropped, and reappeared. A pair of clicking plovers seemed alarmed at some intrusion near the nest, adding their protest to the shrilling sounds of insects coming from every bush and tree.

Breaking into George's reverie I asked him when, in all his days in the bush, he had been the happiest.

'What's wrong with now?' he replied, with an expression of utter contentment upon his face. 'What more could I possibly want than this?' And he gestured all around him and beyond, and as he did so I began to grasp the secret of his philosophy.

Between Sky and Earth

Standing among the time-chiselled rocks high up on the kopje known as Sithwayisi (colloquially called S'twice) which commands a breathtaking view of the Eastern Transvaal Sabie Sand Nature Reserve, I felt very far removed from civilisation's impact. This was wilderness – the raw material out of which civilisation has been fashioned.

With me was Ranger Tim Farrell of Mala-Mala (named after the once plentiful sable antelope, but sadly no longer so), South Africa's exclusive private game camp. It draws visitors from every part of the world by virtue of the unique wealth of animal and bird life contained in a magnificent habitat which, under wise management, is never overcrowded by humans.

There was something special in the air that afternoon, something difficult to define. Was it the faint breeze that touched the high tree-crowns or perhaps the brown-green tinted infinity of woodland and scrub that merged with the horizon? Or the feeling of being suspended between sky and earth, free of tomorrow's uncertainties and yesterday's sorrows?

A whole realm of living creatures was contained around us though we could see and hear but a few from our vantage point. Not even the drone of a jet plane keeping its rendezvous with the world could break the spell of the dynamic wilderness at our feet.

Framed within the bizarre outline of a drifting, fraying cloud, vultures soared upon the thermals. They would continue to patrol their domain, perhaps 3,000 metres above the ground, until the sight of a carcass would draw them back to earth.

A bateleur eagle moved into our line of vision, flipping sideways as it dived at some animal we could not hope to see, even with binoculars. A pair of hawks hovered, then plunged, disappeared and reappeared empty-taloned, then regained height only to repeat the manoeuvre – bringing to mind the story of the crowned eagle, told by ornithologist Terry Oatley. One of Africa's most powerful, skilful aerial hunters, the crowned eagle has evolved an amazing strategy in preying on the vervet monkey which, through generations, has developed protective tactics of its own.

Terry once observed a troop atop their tree hurling abuse at a pair of crowned eagles, believing themselves safe though their enemies were in sight. Suddenly the male eagle swooped down below the canopy, remained out of sight for a few minutes then erupted upwards and, in one powerful wing-rush, snatched a monkey in his talons as he swept past.

Tim Farrell revels in every aspect of his work, even in the feeling of heavy responsibility that rests upon his shoulders – as it does upon every ranger's – when exposing visitors to the element of danger in the bush while exploring the wilds from an open Land-Rover.

Only that evening we had come across two bull elephants which were fortunately giving their full attention to the job in hand, plucking and processing their diet, which at that moment consisted of leaves of a young ebony tree. Tim pointed out saplings of discarded round-leaf kiaat, neatly de-barked by the tuskers – very nut-like in flavour when we sampled a few mouthfuls of what remained.

A bush ranger not only needs to know a great deal about his terrain, from insects to flora, mammals and birds, but he must also be a good judge of character so that the long days far from camp will not cause undue stress to anyone, especially if the visitors haven't had this kind of experience before.

On the return trip to base we met one of those same peaceable bulls on the bend of an upward hill, taking him so unawares that he checked and froze. Tim Farrell didn't rush on and past as another, less bush-seasoned man might have done, but he too checked instantly and waited, looking straight at the elephant, gauging his mood, re-establishing good relations. The tusker relaxed and allowed us to pass without incident. When I looked back over my shoulder he was ambling off into the forest, perhaps towards the river to drink.

There are many kinds of 'rain-trees' in the bushveld, some of which are part of tribal legend. There is the African wattle, known as the huilboom (the tree that weeps), branches of which are planted in the fields along with the first seed sown so that the rains may come. Then there is the weeping boerboom, or tree fuchsia, which grows to a height of more than 10 metres and bears clusters of crimson flowers.

And yet none of these rain-makers actually 'weep' true rain-drops. It is the larvae of certain insects, such as the frog-hopper or spittle-bug and others, that pierce the wood with fine, beak-like mouth-pieces, sucking up sap at such a tremendous speed that it is ejected as froth that 'rains' down from the tree like water.

At dusk Tim took us to a viewpoint above the river where one feels as well as hears the transition from day to night, which comes much more quickly than the merging of night into dawn. Sitting in silence one can pick up myriad sounds that drift across from every level of the vegetation. Are the creatures that inhabit Mala-Mala conscious of one another's existence? Do they react and interact to each other's calls?

Moving rapidly over the river-bed, a raucous drongo mobbed a snake eagle that seemed to be encroaching on its territory, its harsh rebuffs mingling with the metallic click of the plovers and the evensong of the francolins moving off to cover.

The insect day-chorus had ceased and now the night-watch of the scraping legions began to reverberate, bringing down the final curtain of the day, which seemed to have slipped away so quickly yet so gently. Herald of night, the Scops owl, soon began to intone its staccato, purring rhythm as if celebrating the ascent of Venus, first star of evening which cannot come into its own until the last sunset shaft has faded.

Long-crested Eagle

In his book *African Birds of Prey* (Collins, 1975), Leslie Brown, the naturalist and author, states that 'a great many observers in Africa are discouraged from the study of birds of prey by what they regard as a bewildering variety . . . their identification further complicated by plumages that are different in the immature, sub-adult and adult stages and sometimes between male and female.'

How marvellous to be able, as Leslie Brown is, with instant expertise to identify the soaring eagles, to differentiate a harrier from a kite, a buzzard from a hawk, and that under any conditions. Perched and immobile, one bird of prey is definable from another if the plumage is adult and focal points such as legs and beaks, tail feathers and crests, are visible. Often they are not. It's when the view is aerial and the bird is in full flight that wing tips and tail shapes are no more than a blur flashing past overhead. What can be more challenging than to learn to separate one type of feathered aerobat from another, to learn of their habits, their movement patterns, their communication system, their family life.

With Dorothy Hall, a fellow lowvelder, I was recently able to make the acquaintance of a charming eagle foursome, located after weeks of tracking of the adults, which had led us to a gum plantation they had adopted as their home. This type of eagle, the long-crested, isn't too hard to identify, for it carries a long floppy crest on top of its ornate head. Living on the edge of forests rather than deep within the tree-line, this most buzzard-like of the small eagles prefers to raise its young in tall, slender trees, in this case high up in a eucalyptus which had initially – before the advent of the heavy rains – also supported a swarm of wild bees.

I was intrigued by the suggestion that birds of prey at times choose bee-inhabited trees as protection from predators such as snakes and monitor lizards. This gives rise to the question of why eagles themselves should feel immune from bee aggression; or are bee-eagle partnerships merely a matter of chance?

We had spotted the eagle parents before we had entered the plantation, where, within minutes of arrival, we settled into the ground-level sacking hide some 35 metres from the eagles' tree. To my unpractised eye the robust stick-leaf nest at first seemed empty; then suddenly I detected movement just above the nest-rim and within it, amazed to find that the seven-week-old nestling dared to venture forth while the adults were out of range.

For the first time in ten years of bird study Dorothy was able to observe eaglets taking their first steps out of the nest, the larger, older one actually hopping upwards on to a limb where it gave a great wing-flexing display. After this it once again moved back on to the nest and near its younger nest-mate, which as yet did little more than stretch its wings and move about the edge.

For over an hour we waited for the parents' return, our patience finally rewarded by a sudden flutter of black wings as the smaller of the pair, the male, alighted and dropped his morning's catch, probably a rodent, somewhere close to the smaller of the two young. Weeks before Dorothy had witnessed a demonstration of tender, maternal care as the female long-crested eagle had gently and patiently enticed her chick to partake of the meal she presented. Three times she had offered it before it was accepted, using her powerful beak as one might use a long-handled spoon to feed a recalcitrant baby.

The parent eagle had remained on the nest for a few seconds, before flying off to continue the morning's hunt. Not at once, as I had expected, but quite some minutes later, the smaller eaglet began to feed, its elder continuing its morning routine as if nothing of importance had occurred.

When one considers that among many eagles, such as the black variety, nature ordains that the tougher of the chicks destroys the weaker very early on in life, it seemed the more amazing that here, among the long-crests, no conflict arose at the time of feeding. Or had the other bird, perhaps, fed to repletion before we arrived?

Next breeding season may, perhaps, answer this and many other questions, that is, if the same pair return to the same tree where everything will have been prepared for viewing and filming from a tall sacking hide placed much nearer the nest to obtain a complete record of the long-crested eagles at home, *en famille*.

Crane Nocturne

It might well have been the near-full moon that wove a spell over the crowned cranes, as it did over us, as we listened to the voices of the night from the bank of the pool that lies not far from Mlilwane's main camp.

I have seen Swaziland's Game Reserve, Mlilwane, often and in all seasons, yet each time I have been struck anew by the Shangri-la aura of the place, permitting unusual intimacy with the animal inhabitants which one soon learns to identify because Mlilwane is smaller and more accessible than most other sanctuaries in southern Africa.

Only one hundred years ago the same area supported creatures like rhinoceros and vast herds of wildebeest. After the destruction of these great animal communities there followed a period of barren emptiness until Terry Reilly's dream of faunal and floral restoration was realised. What had been farmland became the core of the now internationally famous reserve, a prototype for similar ventures elsewhere in the world at a time when undisturbed retreats are a life-line to sanity.

As Terry Reilly's wife and I waited under the bright moonlit sky for a repeat rendering of the cranes' nocturne, our microphones at the ready, my mind drifted back to the time when Terry's undaunted band battled against enormous odds to relocate such species as ostrich and python, zebra, eland and lynx.

Once the dams had been constructed and indigenous trees planted, migratory birds and water-fowl put down anchorage very quickly. Under the protection of Swaziland's far-sighted, conservation-minded monarch, King Sobhuza II, Mlilwane (named after 'the little hill of fires') thrived, its camp frequented increasingly by visitors, many from far afield.

Because the Reillys ardently believe in the survival of each individual animal, orphans of many kinds find their way to the sanctuary. Some of these are successfully hand-reared and returned to the wild, while others, though free to come and go as they please, remain within reach of the home-base, providing a rare opportunity of studying their ways. Such are the blue and wattled crane pairs that stalk about their domain as if they owned the world, giving one the feeling, and rightly so, that in this sanctuary it is the animal inhabitants that have right of way.

There is a large female warthog with two piglets at foot, which each afternoon insists on sharing the grain that has been spread on the high bird-table; her milk-bar is so voluminous that it is a wonder she manages to balance upright on her hind legs, while her young scuffle and snuffle about her. Woe be to you should you cross her path or aggravate her in any way. At Mlilwane, as nowhere else, you learn to gauge the mood of the animal residents and act accordingly.

An hour and more had passed, though we had not been aware of it, when a single trumpet note rent the valley. Across the pool three crowned cranes suddenly exploded into life, gathering momentum until their voices reached a near-frenzy crescendo. At one moment the sound was akin to a shepherd's reed flute echoing from mountain tops, in the next it shrilled across the water in a rhythmic honking, croaking chorus, accented by a clicking refrain that could have been made by beak or larynx.

What was it that they expressed, with throbbing throats, beaks pointing upwards to the sky? It seemed like a mystical rite celebrating all creation. Was it, perhaps, a genetic relic of an old pattern of behaviour evolved in a time before their terrain and their numbers were drastically curtailed by man's careless encroachment?

The cadenza reached its peak and ceased. Already we had recorded three separate crane cacophonies, the strangest, the most fascinating sound I had yet heard in Africa. The dark shadow of a crocodile suddenly sent the geese, ducks and cranes into a cluck of comment (rather than alarm) and then that, too, gave way to the insect and frog chorus that pervades the night.

Dampness and mist caused a sudden chill and brought us back to reality.

There weren't many hours left for sleep before we would be out again to record the coming of dawn.